DRUG USE

for

GROWN-UPS

ALSO BY CARL HART

High Price

DRUG USE

for

GROWN-UPS

Chasing Liberty
in the Land of Fear

Dr. Carl L. Hart

Penguin Press   *New York*   2021

PENGUIN PRESS
An imprint of Penguin Random House LLC
penguinrandomhouse.com

LIBRARY OF CONGRESS CATALOGING-IN-PUBLICATION DATA

Names: Hart, Carl L., author.
Title: Drug use for grown-ups: chasing liberty in the land of fear / Carl Hart.
Description: New York: Penguin Press, 2021. | Includes bibliographical references
and index.
Identifiers: LCCN 2020004187 (print) | LCCN 2020004188 (ebook) |
ISBN 9781101981641 (hardcover) | ISBN 9781101981658 (ebook)
Subjects: LCSH: Recreational drug use—United States. | Drug abuse—United States. |
Drugs of abuse—United States. | Drug legalization—United States.
Classification: LCC HV5825 .H274 2021 (print) | LCC HV5825 (ebook) |
DDC 362.973—dc23
LC record available at https://lccn.loc.gov/2020004187
LC ebook record available at https://lccn.loc.gov/2020004188

Printed in the United States of America
1   3   5   7   9   10   8   6   4   2

Designed by Cassandra Garruzzo

For Parker and countless other real niggers—
who shielded me from the hit, making it possible for a
hood counterfeit to become mainstream legit.

If you want to get to the heart of the dope problem, legalize it . . . [Prohibition is] a law, in operation, that can only be used against the poor.

James Baldwin

CONTENTS

Author's Note

T his is not a book promoting drug use, nor is it a "how to" book. Today, more than thirty million Americans report using an illegal drug on a regular basis. Drugs don't need an advocate.

I wrote this book to present a more realistic image of the typical drug user: a responsible professional who happens to use drugs in his pursuit of happiness. Also, I wanted to remind the public that no benevolent government should forbid autonomous adults from altering their consciousness, as long as it does not infringe on the rights of others.

I use personal anecdotes and scientific research, my own and others', to dispel drug myths and to illustrate the many potential benefits of responsible drug use. I also share stories that involve other individuals, including relatives and friends. Names and locations have been changed in an effort to protect them from negative repercussions.

After reading this book, I hope you will be less likely to vilify individuals merely because they use drugs. That thinking has led to an incalculable number of deaths and an enormous amount of suffering. I hope you will come away with an appreciation for the prodigious potential good derived from drug use and a deeper understanding of why so many responsible grown-ups engage in this behavior.

Time to Grow Up

If people let government decide which foods they eat
and medicines they take, their bodies will soon be in
as sorry a state as are the souls of those who live
under tyranny.

Thomas Jefferson*

I am an unapologetic drug user. I take drugs as part of my pursuit
of happiness, and they work. I am a happier and better person
because of them. I am also a scientist and a professor of psychology specializing in neuroscience at Columbia University, known for
my work on drug abuse and addiction. It has taken me more than
two decades to come out of the closet about my personal drug use.
Simply put, I have been a coward.

The philosopher John Locke once noted that pursuing happiness

* I recognize that Thomas Jefferson and other revered historical figures enslaved black people.
This was reprehensible even during their time. But the cruel hypocrisy of these individuals'
actions does not negate the noble ideals and vision articulated in their writings. These enshrined principles give us goals to which we continue to aspire.

is "the foundation of liberty."[1] This idea is at the core of the Declaration of Independence, the document that gave birth to our nation. The Declaration asserts that each of us is endowed with certain "unalienable Rights," including "Life, Liberty and the pursuit of Happiness," and that governments are created for the purpose of protecting these rights. The use of drugs in the pursuit of happiness, in my view, is arguably an act that the government is obliged to safeguard.

Why is our government arresting hundreds of thousands of Americans each year for using drugs, for pursuing pleasure, for seeking happiness? The short answer is that it's a very long story. The long answer is the book you are reading. America's drug regime is a monstrous, incoherent mess.

TO GRASP HOW we got here and what we can do about it, I'd like to start by telling you something about my life and work as a "drug abuse" scientist. In the fall of 1999, I landed my dream job, as an assistant professor and researcher at Columbia University's College of Physicians and Surgeons. My research involved giving thousands of doses of drugs, including crack cocaine, marijuana, and methamphetamine, to a range of people in order to study the effects. I believed my work contributed to our understanding of drug addiction. I would be awarded multimillion-dollar grants from the National Institute on Drug Abuse (NIDA) to conduct this research, and I would be invited to serve on some of the most prestigious committees in the area of neuropsychopharmacology. It was a thrilling time.

Twenty years later—twenty years I've spent studying the interactions among the brain, drugs, and behavior and observing how moralizing about drug use is expressed in social policy—my initial excitement has given way to skepticism, cynicism, and disillusionment. When I was a naive graduate student, I believed that I was doing God's work by telling people to stay away from drugs. I believed that the poverty and crime that plagued my childhood community were a direct result of drug use and addiction. I now know that telling people to avoid drugs is no more godly than the Church prohibiting my Catholic wife from using birth control, but it is just as paternalistic, a way of restricting one's freedom and autonomy.

What about the notion that drugs led to poverty and crime in my neighborhood? Well, that is simply an ugly fantasy, an incredibly effective one to be sure. It's effective not only because it is still believed by large segments of the American public but also because it seemingly provides a simple solution to complicated problems faced by poor and desperate people. Many other complex factors are responsible for the turmoil seen in the places of my youth and other communities. But it took me a long time to see that clearly myself. I was too busy for too long being a soldier in the regime, caught up in the cause of "proving" how dangerous drug use is.

I HAD THE COOLEST job in the world. I got people high on a daily basis.

I instructed the twenty-five-year-old white man to light the marijuana cigarette, which was to be smoked through a hollow plastic cigarette holder so that the contents were not visible. He inhaled for five seconds, then held the smoke in his lungs for another ten seconds before exhaling. He repeated this two more times, with a forty-second interval between each puff. We called this our paced-puffing procedure. We used it to standardize, to the best of our ability, the amount of drug inhaled.

Although I couldn't know for sure whether he was getting

placebo or active THC, the major psychoactive ingredient in marijuana, I could tell from his glassy red eyes and the serene smile on his boyish flushed face that he really enjoyed what he had gotten. Nodding slowly and with more bass in his voice than usual, he said, "Yeah, that's it." I could also tell that he was an experienced smoker; it took him only three puffs to suck down nearly three-fourths of the 1 g cigarette. Marijuana smoke now filled the small sterile room.

The smoker, whom I'll call John, was a research participant in one of my studies. And there I was, a young black dreadlocked scientist, trying to conceal the perpetual anxiety I felt about having the strong, distinct smell of marijuana in my hair for another entire workday. I was concerned that as I traveled, on the elevator, from floor to floor or sat in on a lecture or meeting, some judgmental person might think disparagingly, "Typical, dreadlock smoking while at work." Never mind that marijuana has never been my primary drug of choice. Never mind that I had a personal rule, for fear of biasing my results, against using the drug I was currently studying. The year was 2000.

In this particular experiment, I was trying to understand how cannabis affected regular users' brain functioning and behavior. I had received a grant from NIDA to conduct the study. My hard work and commitment were finally paying off. When I started the study, I believed, as did most people, that pot temporarily impairs mental processes such that smokers exhibit memory problems and other cognitive disruptions. There are certainly plenty of anecdotal accounts in line with that view. But, of course, anecdote is not evidence. That's why we do the science. Still, there are even scientific data suggesting that pot temporarily diminishes short-term memory ability in infrequent users.[2] Of course, this is not surprising, because many drugs—alcohol, Ambien, and Xanax, among others—

temporarily disrupt selective mental processes in people who have less experience with that particular drug.[3]

But the negative impact of so-called recreational drug use on the mental functioning of *regular, experienced* users is less clear, at least in the scientific literature. So I was seeking to determine the detrimental cognitive effects of marijuana in people who smoked the drug nearly every day. I wanted to know how they would perform on mental tests after smoking, to establish whether the drug would produce widespread brain dysfunction, even if only temporarily.

John was a typical participant. He smoked multiple joints nearly every day. He was affable, bright, curious, college-educated, and ambitious. He was an artist, an actor between gigs. As a result, he had the time to complete my three-session, outpatient weed study, which paid a couple of hundred dollars. Neither he nor our other research participants fit stereotypical media portrayals of the pothead, who does little besides sitting on the couch, eating Cheetos, and playing video games.

Throughout the experiment, even when under the influence of cannabis, John was lucid and socially appropriate, as were the other participants. No participant failed to show up because they had forgotten the time or day of their scheduled appointment. Not one quit the study because the tests were too difficult or tedious. No one complained that the pot was too weak. And absolutely none of the participants ever became violent. They all complied with our stringent study rules, which imposed demands on their schedules, requiring participants to do considerable planning, to inhibit behaviors that might have been inconsistent with meeting study schedule requirements (e.g., drug use other than marijuana), and to delay immediate gratification.

At the time, I didn't even register the impressive level of responsibility demonstrated by these research participants. I think, despite my best efforts, I mainly saw them as "potheads," "stoners," and all the terms that were inconsistent with the phrase "responsible adults." But I would soon enough find, working with all types of drug users throughout my career, that they have been some of the most responsible people I have ever known.

"Where do you get the pot from?" John asked as he handed the roach back to me. He looked pleasantly surprised to learn that the marijuana he had just smoked was supplied by the federal government. In fact, there is only one pot supplier for researchers in the United States: the NIDA-funded University of Mississippi marijuana-cultivation program.

With a huge smile plastered across his face and a twinkle in his eyes, he said, "Damn, never before have I been so proud of my government." We both laughed, but the joke also took me to a serious place. No one I knew had ever uttered the word "proud" when discussing the U.S. government and pot. Consider the fact that the federal government currently lists marijuana on Schedule I under the federal Controlled Substances Act. This means that the drug is viewed as having "no acceptable medical use in treatment" and is therefore banned in the United States, apart from limited research studies.

This classification is hypocritical, although I only recently came to this conclusion. A plethora of data now demonstrates the medical utility of marijuana. We know—based on research from dozens of scientists, myself included—for example, that the drug stimulates appetite in HIV-positive patients, which could be a lifesaver for someone suffering from AIDS wasting syndrome, and that marijuana is useful in the treatment of neuropathic pain, chronic pain, and spasticity due to multiple sclerosis.[4] The list of conditions for which marijuana has been found to be helpful grows each year.

Therapeutic benefits such as these have compelled citizens to vote repeatedly over the past two decades to legalize medical marijuana at the state level. Today, thirty-three states, plus the territories of Guam and Puerto Rico and the District of Columbia, allow patients to use marijuana for specific medical conditions. In addition, since 1976, the government has supplied pot to a select group of patients, as part of their medical treatment, through the federal medical-marijuana program. And yet federal law still technically forbids the use of marijuana for medical purposes. The inconsistency of federal laws with these initiatives and programs, and with the increasing number of studies demonstrating the medical usefulness of the substance, highlights our government's hypocrisy and undoubtedly undermines people's trust in the government when it comes to regulating other drugs as well.

Not only has the trust of governmental regulatory agencies been eroded as a result of their handling of specific drugs, a growing number of people have begun to question the objectivity of government-funded scientists who study drugs. Consider frequent statements made by some of these scientists, including Dr. Nora Volkow, director of NIDA, which emphasize the possible neurological and psychiatric dangers of drug use—cannabis included—while virtually ignoring these drugs' potential medicinal or other beneficial effects.

Nora and other scientists have been quick to caution that pot, for example, is a "gateway" drug to harder substances, but they never mention the more than half a million Americans who are arrested each year mainly for simple possession of the drug, to say nothing of the shameful racial discrimination in marijuana arrests. At the state level, black people are four times more likely to be arrested for marijuana than their white counterparts.[5] At the federal level, Hispanics represent three-fourths of the individuals arrested for marijuana violations.[6] This is despite the fact that blacks, Hispanics, and whites all use the drug at similar rates,[7] and they all tend to purchase the drug from individuals within their racial groups.[8]

I would learn later in my career that the marijuana gateway theory grossly overstates the evidence by confusing correlation with causation. It's true that most cocaine and heroin users started out using marijuana first. But the vast majority of pot smokers never go on to use so-called harder drugs. To say marijuana is a "gateway" to "harder" drugs is baseless: correlation, a mere link between factors, does not mean that one factor is the cause of another.

I myself had long been guilty of focusing almost exclusively on the harmful effects produced by drugs, including marijuana. In the above experiment, for example, I didn't even consider the fact that cannabis might not produce negative effects on mental performance, let alone that it might produce positive ones. In June 2000, I was invited to give a talk at a meeting of the Behavioral Pharmacology Society. My study had not yet been completed, but preliminary data were showing that the drug produced virtually no disruptive effects on the complex mental abilities (e.g., reasoning and abstraction) of regular users and that it even improved performance on a test of vigilance. And in terms of mood, the drug produced euphoria and pleasurable feelings.

Never mind: at the conclusion of my talk, I virtually ignored any beneficial effects and speculated that perhaps had I given participants multiple marijuana cigarettes prior to testing their mental functioning, I would have observed more cognitive disruption. Dr. Jack Bergman, a Harvard Medical School–based psychobiologist, asked me a reasonable follow-up question: "Is it possible that marijuana, at euphoric doses, is without effect on cognitive flexibility, mental calculation, and reasoning, at least in this group of subjects?"

I was so utterly focused on the harmful effects of drugs that I couldn't see this as a possibility, even though it was exactly what the data were showing. Stumped, I managed to babble on about the possibility of including more complicated test measures in future studies.

Jack's question would continue to nag at me. More and more, I came to realize that drug-abuse scientists, especially government-funded ones, focus almost exclusively on the detrimental effects of drugs, even though these are, in fact, a minority of effects. This has had a damning impact on how so-called recreational drugs are regulated and inevitably on your own decision as to whether or not to partake of them.

Here's the bottom line: over my more than twenty-five-year career, I have discovered that most drug-use scenarios cause little or no harm and that some responsible drug-use scenarios are actually beneficial for human health and functioning. Even "recreational" drugs can and do improve day-to-day living. Several large research studies have shown that moderate alcohol consumption, for example, is associated with decreased risk of stroke and heart disease, the top killers in the United States each year.[9] As you will discover, a number of beneficial effects have been observed with other drugs as well. From my own experience—the combination of my scientific work and my personal drug use, I have learned that recreational drugs can be used safely to enhance many vital human activities.

WITH SOME TREPIDATION, I chart in this book for the first time my awakening as a citizen-scientist trying to make people aware of these facts. I also describe my struggles to convince other drug-abuse researchers that we operate under some important biases, which in some cases are more damaging than the drugs themselves and prevent us from exploring new treatments and healthier, humane policies. I provide detailed strategies that you, as a *responsible, adult* drug user, can employ in order to enhance positive drug effects, while minimizing negative ones. These are the same strategies that I use in my government-funded research to keep research participants safe.

A point I need to emphasize here is that this is a book for grownups. By that I mean autonomous, responsible, well-functioning, healthy adults. These individuals meet their parental, occupational, and social responsibilities; their drug use is well planned in order to minimize any disruptions of life activities. They get ample sleep, eat nutritiously, and exercise regularly. They don't put themselves or others in physically dangerous situations as a result of their drug use. These are all grown-up activities.

Growing up is difficult and it's not guaranteed. In other words, neither this book nor drug use is for everyone. They are for those who have managed to grow up.

I recognize that people with mental illness and those experiencing acute emotional crises (e.g., the death of a loved one or a divorce) may also be interested in the ideas expressed within these pages. But because people with specific mental illnesses and those in crisis are at greater risk for experiencing negative drug-related effects, it would be irresponsible of me to encourage use by these groups without detailing each caveat associated with any particular substance and psychiatric disorder. Frankly, that analysis is beyond the scope of this book.

A related issue is drug addiction. *Drug Use for Grown-Ups* is unapologetically not about addiction. But because I use the terms *addict* and *addiction* throughout, it's incumbent upon me to clearly define them. Simply knowing that a person uses a drug, even regularly, does not provide enough information to tell whether that person is "addicted." It doesn't even mean that the person has a drug problem. To meet the most widely accepted definition of addiction— the one in psychiatry's *Diagnostic and Statistical Manual of Mental Disorders, 5th Edition (DSM-5)*—a person must be distressed by their drug use. In addition, the individual's drug use must interfere with important life functions, such as parenting, work, and intimate relationships. This use must take up a great deal of time and mental energy and must persist in the face of repeated attempts to stop or cut back. Other symptoms that the person may experience include needing more of the drug to get the same effect (tolerance) and suffering withdrawal symptoms if use suddenly ceases.

My use of the term *addiction* throughout this book is interchangeable with *DSM-5*'s *Substance Use Disorder*, which always means problematic use of the sort that interferes with functioning—not just ingesting a substance regularly.

Too often, the conversation about recreational drug use is hijacked by peddlers of pathology as if addiction is inevitable for everyone who takes drugs. It is not. Seventy percent or more of drug users—whether they use alcohol, cocaine, prescription medications, or other drugs— do *not* meet the criteria for drug addiction. Indeed, research shows repeatedly that such issues affect only 10 to 30 percent of those who use even the most stigmatized drugs, such as heroin and methamphetamine.[10] This observation highlights two important points. The first is society's flagrant, disproportionate focus on addiction when discussing drugs. Addiction represents a minority of drug effects, but it receives almost all the attention, certainly the media attention. Think about that. Have you ever read a newspaper article or seen a film about heroin that didn't focus on addiction? Imagine if you were interested in learning more about cars or driving and could only find information about car crashes or information about how to repair a car after a crash. That would be ridiculous.

Another related point is this: if most users of a particular drug do not become addicted, then we cannot blame the drug for causing drug addiction. It would be like blaming food for food addiction. Can you imagine us waging a war on cheesecake or steak? You've seen the histrionic headlines that blame specific drugs for their extraordinary "addictive powers," as if certain drugs have magical qualities. Drugs are inert substances. The evidence tells us that we must look beyond the drug itself when trying to help people with drug addiction. In fact, regarding the relatively small percentage of individuals who do become addicted, co-occurring psychiatric disorders—such as excessive anxiety, depression, and schizophrenia—and socioeconomic factors—such as resource-deprived communities and un- and underemployment—account for a substantial proportion of these addictions.[11]

I recognize, too, that nowadays it's nearly impossible to engage in a discussion about drugs without addressing their purported negative impact on the brain. You will discover within these pages that scientists have frequently overinterpreted and distorted many of these effects. Adding to the problem, misrepresented "brain findings" are then amplified by less than careful media coverage. By looking critically beyond the pretty pictures produced by brain imaging, I will challenge the notion that recreational drugs cause brain dysfunctions. You will see that the sexy images so frequently touted by some neuroscientists rarely show any actual data, but this doesn't temper the unsubstantiated claims made about the brain-damaging effects produced by drugs. This irresponsible behavior, you will see, has contributed to inappropriate drug policies that have led to racial discrimination, group marginalization, and preventable deaths.

A broader argument I make within these pages is that adults should be permitted the legal right to sell, purchase, and use recreational drugs of their choice, just as they have the rights to engage in consensual sexual behaviors, drive automobiles, and even purchase and use guns. Of course, all these activities carry some level of risk, including death. But rather than banning sex, cars, or guns, we have implemented age and competence requirements as well as other safety strategies, strategies that minimize harms and enhance positive features associated with these activities. This is already done, of course, with the widely used recreational drug known as alcohol. After reading this book, you will, I hope, come to the inescapable conclusion that the same should be done with other recreational drugs.

Recreational drug use is an activity engaged in by millions of closeted adults around the globe. Now that I have learned that taking drugs to alter one's state of being isn't as dangerous as I had been taught, I share my story in an effort to encourage others, especially successful professionals who are less at risk than people on the margins of society, to get out of the closet about their own drug use. If they did so, more people would see that there are far more respectable drug users than our criminal-justice regime and popular culture would have us know.

Media coverage of the current so-called opioid crisis is but one clear example of the pervasive spread of misinformation about drugs and the people who partake of them. This type of coverage has made it damn near impossible for rational adults to acknowledge publicly their recreational opioid use. According to the lore, one must be in excruciating pain, mentally ill, or extremely troubled to use opioids, because any use is said to be accompanied by a high risk of addiction, overdose, and death. The same was said about methamphetamine in the early 2000s and crack cocaine in the late 1980s. I'm embarrassed to say that I learned that such statements simply aren't true not from critically analyzing my research data, but from my own personal drug use.

Heroin and other opioids, such as oxycodone and morphine, bring me pleasurable calmness, just as alcohol may function for the drinker subjected to uncomfortable social settings. Opioids are outstanding pleasure producers; I am now entering my fifth year as a regular heroin user. I do not have a drug-use problem. Never have. Each day, I meet my parental, personal, and professional responsibilities. I pay my taxes, serve as a volunteer in my community on a regular basis, and contribute to the global community as an informed and engaged citizen. I am better for my drug use.

But I am also a parent of a teenager and young adults. So, you may ask, how could I, in good conscience, admit to using some of our most vilified drugs, especially now that the country is experiencing an opioid "crisis"? Am I not concerned what my children will think? Am I not concerned that publicly acknowledging my own drug use will increase the likelihood of my own children using drugs? Also, not least, am I not breaking the law by using heroin?

The answers to these questions lie in my story and in the science, both of which speak to how society is constantly misled about drugs and how this leads not only to countless preventable deaths but also to policies that force adults to behave like children and to social conventions that place ridiculous moratoriums on the use of mind-altering drugs for pleasure. By exploring the myths and social forces that shape our views on drugs and policy, we can tear away the misinformation that actually drives so-called drug crises and get down to the vital business of pursuing happiness.

1

The War on Us:
How We Got in This Mess

The degree of civilization in a society
can be judged by entering its prisons.

Fyodor Dostoyevsky

Hey Carl! Caaarrrl!" a woman's voice yelled from behind, as I walked across the picturesque Columbia campus on a crisp February day. I was headed to Sing Sing Correctional Facility to teach a course on drugs and human behavior. Every Friday evening during the spring semester, I made the hour-and-a-half journey from the university to prison, carrying with me a host of conflicting emotions, ranging from a great deal of sadness to pride to complicity.

I also carried my noise-canceling headphones, and I didn't initially hear the woman's call because I had already donned them. I was lost in the rich bass-baritone voice of Isaac Hayes singing his

1971 song "Soulsville." In his way, Hayes was calling out the key forces that hampered black men's economic mobility fifty years back. "Any kind of job is hard to find," he sang. It took me back. Growing up, I remember my mom stressing, "If you ain't got no job, you ain't no man." The similarities between the conditions that Hayes described *then* and those faced by my students at Sing Sing decades later are so painful, in part, because they are so obvious and fixable.

The woman shouting my name caught up with me and then had my full attention. It was Ruth, a fellow Columbia faculty member I respected, whom I had known for at least a decade. "I went to Sing Sing last night!" she proclaimed, a giddy smile stretching from ear to ear. It wasn't a polite, half-hearted grin; it was authentic, spontaneous, and bubbly, even euphoric. She was extremely pleased with herself for having gone to Sing Sing.

Ruth explained that she had taught her first lecture there the previous evening, and she wanted to share with me her excitement. "I had a great time!" she exclaimed. I knew she didn't mean it this way, but it came across as if she were describing an experience at summer camp.

I really wanted to match her joy but couldn't. In the three years I had taught at Sing Sing, I couldn't recall ever thinking, "I had a great time" following a lecture in that dehumanizing space. This isn't to say I didn't have a sense of subversive accomplishment from teaching my students how to think, how to identify the hypocrisy and double standards that contribute to their subjugation. But consider what happens when you first arrive for the evening. After making sure you have no electronic devices or anything other than your ID on you, a guard, who greets you with icy indifference, makes you wait outside, sometimes in the bitter cold, sometimes while slowly finishing her meal, sometimes slowly inspecting the same ID she has seen many times before.

Eventually you are summoned to enter the building and ordered to remove your shoes, belt, and other items for inspection. This is followed by passage through a metal detector. All is scoped by the watchful eyes of the chief white overseers: governor Andrew Cuomo, commissioner Anthony Annucci, and warden Michael Capra. Large photos of these men hang on the wall just below huge lettering that reads:

WELCOME TO SING SING CORRECTIONAL FACILITY

WE ARE A TEAM OF ONE

COMMITTED TO SERVE WITH HONOR, INTEGRITY, AND PROFESSIONALISM

Once cleared, you are locked in a cell about ten by ten feet, with three worn wooden benches, a pay phone, and a wooden suggestion box hanging on the drab brick wall. On the opposite wall, a bulletin board contains materials celebrating Black History Month and inspirational quotes from people such as Martin Luther King Jr. You can remain in this cage for as little as fifteen minutes or for close to an hour before taking a ninety-second bus ride to the classroom building. The wait time is up to the discretion of the guard in charge.

Almost all the other instructors were women and white; the students were male and mostly black. I buried my head in lecture notes, pretending not to be tuned into the conversations taking place around me. Most instructors were chatty and appeared remarkably untroubled about being locked in a cage at Sing Sing. Not me. I was perpetually uneasy, fearing the day when some guard would say I "fit the description" and must remain locked behind bars.

On my third trip to Sing Sing, I had my heart broken. Walking into the classroom, I was greeted by an unexpected voice: "What's up, cuz?" The words were followed by some dap and a warm embrace. It was Robert, my first cousin Sandra's eldest child. He was wearing green trousers and a green sweatshirt, the standard Sing Sing inmate uniform. I was shocked. I didn't even know he was locked up, let alone here.

I hadn't seen Rob since we were children but had a vague memory of family stories of his chaotic upbringing. He and his siblings had been removed from their mother's custody before he was a teenager, and things didn't get better from there.

As it turned out, Rob had killed someone, a rival drug dealer, and was now doing twenty-five years to life. As he told it, he had simply beat the man to the punch: his rival had set in motion a plot to kill him because Rob had recently taken over much of his territory.

As Rob walked away, I felt deflated and wondered how I was going to get through the next two hours of teaching. I stood there, alone, searching for answers in that empty, cold, quiet classroom. I reminded myself of my sense of obligation to teach there, my feeling that it is a civic duty. I recalled the sincere pride in the eyes of a student when he told me he'd never met an author prior to taking my course, let alone a black author. I deeply admired my students' enthusiasm and the intellectually passionate way they tackled the curriculum, not least because some of them had a personal stake in the subject. A number were serving time for drug-related offenses.

THE WAR ON US

It's impossible to talk about drugs without addressing the elephant in the room—or, more aptly, the albatross around the necks of specific groups—*the war on drugs*. The ostensible goal of this U.S. government–led campaign is to eradicate certain psychoactive drugs. Today, the American taxpayer spends approximately $35 billion each year fighting this war.[1] Yet the drugs in question remain as plentiful, if not more so, than they were in 1981, when the sum total of America's annual drug-control budget was a mere $1.5 billion.[2] What has changed is that now, each year, tens of thousands of Americans die from drug-related overdoses. The popular notion holds that opioids are the primary culprit, but as we will see, it's not that simple.

GIVEN SOCIETY'S RETURN on the twentyfold increase in our drug-control budget, we could reasonably conclude that the war on drugs has been a complete failure.

It has not. Otherwise this country would not have continued to perpetrate this war decade after decade after decade. True, the war on drugs has not succeeded in the impossible and unrealistic task of ridding society of recreational drugs. Only children and naive adults honestly believe that this was an actual or achievable goal. A vital but unstated aim of the drug war is to shore up the budgets of law-enforcement and prison authorities, as well as such parasitic organizations as drug-treatment centers and urine drug-testing outfits. Law-enforcement entities receive the bulk of federal drug-war dollars.

Here's an example of how it goes down: entire specialty police units are deployed to poor, usually black or brown neighborhoods, making excessive drug arrests and subjecting targeted communities to dehumanizing treatment. The argument that these communities are exposed to "enhanced police presence" because residents ask for it is either naive or disingenuous; these are the same residents who have asked for, and in fact demanded repeatedly, better schools, more jobs, and an end to police brutality, as well as placed a long list of other reasonable requests.

The bottom line is simple: more drug arrests equate to more overtime, more "throwaway people" in prison, and bigger budgets. These practices ensure job security for a select few, including law-enforcement personnel and prison authorities. The war on drugs has been a financial boon for these individuals, as well as for certain regions that are dependent upon the prison economy. Most prisons in the state of New York, for example, are located in rural, white communities. The prison is usually the major employer in the area. And because a large proportion of the inmates come from areas many miles away, their visiting loved ones have to frequent local restaurants, hotels, and other local businesses. And in states such as Pennsylvania, an incarcerated inmate is counted as a resident of the jurisdiction where their prison is located for purposes of allocating financial resources from the state, a grotesque new twist on the U.S. Constitution's original 3/5th clause. It's not difficult to see how the war on drugs has been hugely beneficial for some.

Along the way, however, specific minority communities have been devastated. Complex economic and social forces are routinely reduced to "drug problems," and resources are directed to those in law enforcement rather than to neighborhoods' real needs, such as job creation, better education, or lifesaving drug services (discussed in Chapter 3). This is how every "drug crisis" has played out up through today. In essence, the war on drugs is not a war on *drugs*; it's a war on *us*.

A CHANGE IS GONNA COME . . . OR IT WON'T

Back in the classroom in Sing Sing, we were in the middle of a heated discussion about how the drug war would be carried out during the current "opioid crisis." "You never let a serious crisis go to waste," Rahm Emanuel once remarked, "and what I mean by that [is] it's an opportunity to do things you think you could not do before." In line with this view, perhaps the opioid situation could actually provide an opportunity to push for meaningful movement toward regulating all drugs, just as we effectively regulate alcohol or cannabis (in some states). Regulation would certainly reduce the number of deaths caused by contaminated drugs. It would also decrease drug-related arrests and would permit adults freedom to make reasonable decisions about their own drug use. On the other hand, perhaps the current crisis will make the situation even worse—by leading to interventions that further restrict individual liberty and providing another reason to arrest certain Americans at high rates—without helping to solve the purported problem.

"I hate to say it," Hakeem reluctantly piped up, "but the opioid crisis is ultimately a good thing." He speculated that because white Americans are seen as the primary users, opioid use—and other drug use by extension—would no longer be treated as a crime. Instead, he predicted, it would be treated as a health problem, an approach that would be beneficial for everyone regardless of race.

Several other students agreed. They highlighted the public perception of large numbers of white Americans experiencing problems related to opioid use, including fatal overdose and addiction. This perception, some felt, has engendered unprecedented sympathy from the broader community for drug users. In 2017, Donald

Trump even proclaimed the problem a national emergency. His announcement appeared to consolidate a definite shift in the way the country views certain drug users. They are now patients in need of our help and understanding, rather than criminals deserving scorn and incarceration.

Signs of this shift were evident as early as January 2014. Then governor of Vermont, Peter Shumlin, devoted his entire State of the State address that year to the "heroin crisis" and urged his overwhelmingly white electorate to deal with addiction "as a public health crisis, providing treatment and support, rather than simply doling out punishment, claiming victory and moving on to our next conviction."[3] Politicians around the country from both parties have echoed these sentiments, and in 2018 Congress passed a multibillion-dollar bipartisan bill aimed at curbing opioid-related problems (H.R.6).

What looks like a radical shift to a more compassionate drug policy—one that favors treatment (and other support) over incarceration—has encouraged several of my students, as well as many others, to hope that we are entering an age in which there will be far fewer drug-related arrests and deaths than there were in previous decades.

But a contingent of students was not so optimistic. Mike, for one, gave a firm rebuttal to Hakeem's comments. "Nah, I disagree," he snapped. "White still means victim, and black and Hispanic still mean addict and criminal."

After several minutes of back and forth, the students wanted to hear from me. They wanted to know on which side of the debate I stood.

"Of course I support an approach that favors treatment over incarceration," I told them. But these aren't the only options. There are a range of other possibilities, including the removal of criminal sanctions for adults who consume drugs responsibly. At the moment, simple possession of any controlled substance can land a first-time offender in prison for up to one year. On top of this, the person must pay a fine of no less than $1,000. The law gets considerably crueler with subsequent violations or when trafficking or manufacturing are the charges. So offering treatment over incarceration is the bare minimum that we should do, certainly as it relates to dealing with individuals who are struggling with drug addiction. But it is not, historically, what we have done for *all* our citizens.

I asked the students to recall a previous lecture when I discussed the "crack crisis" of the late 1980s. "Can you imagine," I asked rhetorically, "Governor George Wallace of Alabama urging his voters to view crack use as a health crisis?" Back then, even northern liberals—both black and white—were calling for foolish and draconian measures to deal with perceived users and sellers of crack. New York governor Mario Cuomo lobbied for life sentences for anyone caught selling crack, in amounts of as little as fifty dollars, while Harlem congressman Charles Rangel advocated for the deployment of military personnel and equipment to rid cities of the drug. The fear of crack and its sellers and users caused mass hysteria. As a result, in 1986 and 1988, Congress passed and extended the infamous Anti–Drug Abuse Act (a.k.a. crack-powder laws), setting penalties that were one hundred times harsher for crack than for powder-cocaine convictions.

The perceived users and sellers of crack were black, young, and menacing, and the public contempt expressed toward this group was intense, visceral, and widely encouraged. In reality, most crack users were white, and most drug users bought their drugs from dealers within their own racial group.[4] By 1992, though, more than

90 percent of those sentenced under the harsh crack-powder laws were black.[5] They were required to serve a minimum prison sentence of at least five years for small amounts of crack. Under the 1988 law, even first-time offenders were subjected to this stiff penalty for simply possessing crack cocaine. No other drug-law violation resulted in such harsh punishment for first offenders.

To the extent that the use of crack by whites was acknowledged, media reports sympathetically detailed the plight of white middle-class crack users. Crack was understood to be a tool for managing stressful professional lifestyles.

For whites afflicted with crack addiction, medical experts extolled the effectiveness of treatment. Any law-enforcement perspective was conspicuous by its absence. Public service announcements (PSAs), geared toward middle-class crack users, encouraged sympathy and not judgment. Sound familiar?

This pattern of racial differentiation—one drug policy for white users and another for black users—followed the format carried out during the heroin crisis of the late 1960s. In the media, the face of the heroin addict was black, a destitute person engaged in repetitive petty crimes to feed his or her habit. A popular solution was to lock up these users. New York State's infamous 1973 Rockefeller drug laws exemplified this perspective. This legislation created mandatory-minimum prison sentences of fifteen years to life for possession of small amounts of heroin or other drugs. More than 90 percent of those convicted under the Rockefeller laws were black or Latino, even though they represented a minority of drug users.[6]

This punitive approach to black heroin users coincided with a massive expansion of methadone maintenance programs that benefited large numbers of white "patients," including addicted soldiers returning from the Vietnam War.[7] Even President Nixon praised methadone "as a useful tool in the work of rehabilitating heroin addicts," one that "ought to be available to those who must do this work."[8]

One feature of methadone programs that was viewed as a drawback was the requirement that the drug be administered through health clinics or hospitals. This meant that patients had to attend the clinic daily in order to receive the medication. This requirement presented an inconvenience for some patients, especially those with jobs and demanding schedules. Also, the fact that patients were required to stand in line outside the clinic in order to receive the medication was viewed as stigmatizing, a form of public shaming. So, in 1971, New York City mayor John Lindsay pushed for the use of private physicians to distribute methadone to a select group of middle-class and insured patients, leaving mostly poor people to stand in those lines, locking in the public face of methadone users.[9]

This characteristically American pattern of cognitive flexibility on drug policy, with harsh penalties for some and sympathetic treatment for others, has a long history.

RACISM

I have been well aware of the differential response, based on race, to drug users for quite some time. A recent conversation I had with my friend Abby let me know that others also are tuned in. Abby is white, old enough to be retired, and financially well off; she's also a lifelong pot smoker. On this particular evening, we were having dinner in a location outside her home state, a place where recreational marijuana is still banned. We both had just arrived in town only hours before. So when she pulled out her marijuana-filled pipe to smoke, I expressed surprise at how quickly she'd managed to get her drug of choice. Abby told me she had brought the drug with her on the plane and that this has been her habit for many years. "Fuck, I'd be too petrified to do that," I said. "Aren't you?" She replied, "Carl, look at me . . . I'm white, old, and rich. Who's gonna fuck with me?"

"Touché" was all I could say. She was right. She succinctly described her white privilege in the drug context. "More power to her," I thought. She recognized and exercised her privilege. Nothing wrong with that. Besides, what societal benefit is gained from arresting Abby for possessing personal-use amounts of marijuana? Absolutely none. She's a responsible consumer and an upstanding citizen, a pillar of her community.

It would be ideal if we as a society extended this white privilege to all our citizens. Unfortunately, it doesn't work that way, especially when it comes to the enforcement of drug laws. In fact, the privileges afforded to some are acquired at the expense of others. This phenomenon can be viewed as the inverse of white privilege—racial discrimination or racism. When I use these terms here, I simply mean an action that results in disproportionately unjust or unfair treatment of persons from a specific racial group. Malicious intent is not required. What is required is that the treatment be unjust or unfair and that such injustice is disproportionately experienced by at least one racial group.[10]

Black people are much more likely than their white counterparts to be arrested for drugs, even though both groups use and sell drugs at similar rates.[11] Not only is this wrong, but it has created a situation where law-enforcement agencies suspect damn near every black person as a drug trafficker.

I travel extensively but dread having to go through customs in some countries because invariably I am grilled about whether I'm carrying drugs. I recall once traveling through the Toronto airport and being taken to a back room for further questioning, supposedly about my visit. I explained that I was headed to Thunder Bay to give a public lecture. But this wasn't good enough. More interrogation followed, along with an examination of the contents on my computer. After what seemed to be an inordinately long period of time, I grew impatient and said, "Look, I'm a scientist . . . and a professor . . . and an author . . . Here's a copy of one of my books." The incredulous look and impish smile plastered across the face of this white Canadian customs officer told me she wasn't impressed. "Just because you wrote a book," she said, "doesn't mean you're not a drug dealer." Being a black man traveling from one country to another was enough for me to be a suspected drug dealer, no matter that all the evidence I presented was consistent with whom I claimed to be.

I don't know if this officer is a racist or not. I suspect she wouldn't consider herself one; few people do. Still, these types of experiences have made me think a lot more deeply about what constitutes racism.

It's easy to classify as racists people who acknowledge they willfully perpetrate racial discrimination. But who's dumb enough to admit to being a racist, excluding self-proclaimed white supremacists?

And what about those who unknowingly participate in racial discrimination? The cop who was "just doing his job"? Or the well-meaning legislators who voted for the crack-powder laws that were enforced in a racially discriminatory manner? Are these people racist? From my perspective, this determination can only be made by assessing their response to reasonable evidence that their actions contribute to racial discrimination. If the cop and lawmaker unwittingly participated in racial discrimination but changed their behavior when the discrimination was brought to their attention, it would be inappropriate to label them as racists. We all make mistakes. On the other hand, if these individuals fail to take action after being presented with such evidence, then the label "racist" is appropriate.

The key is to keep the focus on people's actions, on their behaviors, rather than to speculate about their motives. Trying to determine what's in a person's head or heart is a pointless distraction. It's impossible to know, for certain, the heart's inner secrets.

Similarly, it is not helpful to focus on "implicit bias" because such unconscious attitudes may or may not play a role in the act of racial discrimination. In other words, simply having an implicit bias does not mean a person will inevitably act on this bias in a racially discriminatory manner. Nor does it mean a specific act of racial discrimination was due to such bias. Placing attention on implicit bias—on *a person's thoughts*, rather than on *that person's harmful acts*—tends to obfuscate the issue. Such emphasis is frequently a device used to avoid addressing head-on what is obvious racism, such as that which occurs in drug-law enforcement.

WHY ARE DRUGS BANNED IN THE FIRST PLACE?

On December 10, 1986, James Baldwin was the keynote speaker at the National Press Club luncheon. Just forty-four days earlier, the Anti–Drug Abuse Act went into effect. Baldwin took the opportunity to criticize the new legislation, referring to it as "a bad law." He predicted that it would exacerbate racial discrimination and "only be used against the poor." He also urged black politicians, specifically, to push for drug legalization on behalf of their constituents. Sixteen of the twenty Congressional Black Caucus members had voted in support of the tough new law.[12]

Back then, I was an airman in the U.S. Air Force stationed at Royal Air Force Fairford in Gloucestershire, England. I was a part of the base's security police unit. I hadn't always been a cop, nor did I want to be one. But on April 14, 1986, our country bombed Libya in retaliation for Libyan-sponsored terrorism against American soldiers and citizens. The KC-135 planes that provided aerial refueling for the bombers came from our base, so we were on high alert for counterattacks.

As part of the enhanced base-security measures, I was selected, much to my chagrin, to be a security police augmentee. In my new job, I patrolled the base with an M16 rifle, sometimes for sixteen hours straight. I hated this duty. But I did as I was told because I had taken an oath to obey the orders of my superiors and because I had vowed to support and defend the Constitution against all enemies, foreign and domestic. I didn't think of myself as being particularly patriotic. I was only doing what was right. It was right just as not killing another human being is right. It was right just as not lying is right. It was right just as not taking drugs is right. It was right and it was simple.

Baldwin's remarks, to my way of thinking, were wrong. I sat in quiet disbelief, listening intensely as he made his case. His suggestion that cops would seize the opportunity—provided by the new statute—to arrest black people selectively was hard to take. "If people don't use or sell drugs," I thought to myself, "then they won't get arrested." Even though, by this time in my life, I had been stopped by the police more than once for no other reason than my skin color, I was still too naive to appreciate fully that certain communities were overpoliced and subjected to unfair treatment by the police.

Baldwin's dispassionate, nonjudgmental comments on both drugs and legalization were different from the dominant public narrative. That he didn't condemn drugs seemed strange. His views were disconcerting. They certainly weren't informed by the countless PSAs that cluttered the air with powerful antidrug warnings delivered by celebrities. "Smoking crack is like putting a gun in your mouth and pulling the trigger" was the frightening message in a PSA that left an indelible impression on me. I worried that Baldwin's recommendations would lead to more drugs and chaos in resource-poor neighborhoods like the one from which I came.

Baldwin's views on drugs seemed irresponsible. I was perplexed and disappointed. He was one of the few thinkers I truly revered. His writings helped me to see that white Americans, as a group, were not my enemy, even if, on occasion, a few frustrated the fuck out of me. Baldwin's words expressed this relationship with our white brothas and sistas eloquently: "I never really managed to hate white people, though god knows I have often wished to murder more than one or two."[13]

I now know that Baldwin was right about drugs, just as he was right about so many other important issues. Enforcement of the crack-powder laws did, in fact, lead to rampant racial discrimination in arrests, prosecutions, and convictions. The effects of this abhorrent practice continue to reverberate to this very day. It would take me more than a decade to become aware of this injustice, even though several of my own friends and relatives had been arrested and had served time for violating this law.

This realization made me rethink my views about drugs and their regulation. I'm embarrassed to admit it now, but I once wholeheartedly believed that drugs destroyed certain black communities. This despite the fact that over the same period of time, I attended an inordinate number of social events hosted by white colleagues. The setting for such events was usually in white communities. And almost without fail, psychoactive substances—both legal and illegal—served as social lubricants. The availability of drugs abounds. I assure you, though, drugs have not destroyed these white people or their communities. The folks to whom I am referring are some of the most responsible and respectable people I know. They are scientists, politicians, educators, activists, entrepreneurs, artists, media personalities, and more. They are your children, your siblings, your parents, your grandparents. They are you . . . me. And they are drug users, albeit mostly closeted drug users.

THE DECLARATION OF INDEPENDENCE

So serving in the military didn't cure me of my naive inherited wisdom about drugs. But I am grateful for having served because it was there that I first developed a profound appreciation for the three documents that gave birth to our nation—the Declaration of Independence, the Constitution, and the Bill of Rights.

Of the three, I am inspired most by the powerful concepts articulated in the Declaration of Independence. Even though it is not law, the Declaration is the foundation upon which American democracy was built. This document guarantees each citizen three birthrights—"life, liberty and the pursuit of happiness"—that can't be taken away. It proclaims each person's right to live as they see fit, as long as they do not interfere with others' ability to do the same. And it declares that governments are created "to secure these rights," not to restrict them.

For more than twenty-five years, I've studied drugs, trying to understand how they affect the brain, mood, and behavior. I've also written extensively on drug policy. It took me many years to see that the concepts expressed in the Declaration have profound ramifications for drug policy. In fact, the Declaration defends an individual's right to use drugs. Clearly, many people consume psychoactive substances "in the pursuit of happiness," a right the government was established to secure, to protect. So why then is our current government arresting one million Americans each year for possessing drugs? Why are so many drug users hiding in the closet? This reality does not align with the spirit of the Declaration.

Nor does it align with how drug use was handled for most of American history. From the country's inception through the early years of the twentieth century, Americans were free to alter their consciousnesses with the substances of their choice. A range of over-the-counter concoctions containing alcohol, cocaine, opioids, and other psychoactive drugs was readily available. Opium was the sought-after constituent in several general feel-good remedies, and cocaine served as the most important ingredient in tonics such as *Coca Cola*.[14]

Upstanding citizens openly used drugs to feel nice, to alter their consciousness. Thomas Jefferson, the author of the Declaration of Independence, was a long-term, avid drug user. He particularly appreciated the opium-based drugs for their mind-altering as well as medicinal effects.[15] Sigmund Freud was perhaps the best-known proponent of cocaine use. He himself used it to improve his mood and increase his energy. Everyday people, too, enjoyed drugs such as cocaine and opioids without shame. In fact, the typical consumers of opioids were middle-aged white women. They bought opium or morphine from the local store and used the substances with few problems.[16] This "allowing adults to be adults" attitude was soon to change.

After the American Civil War, Chinese workers were brought into the United States to help build the railroads. Some of them brought with them the practice of smoking opium. Opium dens, which were usually run by the Chinese (in China, the drug could be obtained and consumed freely), were increasingly frequented by white Americans. This intermingling inspired racial fear; numerous media reports followed claiming that opium use was widespread and that good, young white people were being corrupted in the dens. This excerpt from an 1882 report was typical: "The practice spread rapidly . . . Many women and young girls, as also young men of respectable family, were being induced to visit the dens, where they were ruined morally and otherwise."[17]

Similarly, use of cocaine by black day laborers and other blue-collar workers was initially encouraged, as long as the use was in the service of accomplishing work tasks for whites. But then the situation changed as whites discovered that blacks, too, enjoyed cocaine recreationally for its euphoria- and confidence-inducing effects. Use by blacks was increasingly reported in a manner designed to evoke fear among the white majority. Countless articles exaggerated both the extent to which cocaine was used by blacks and the connection between their use of the drug and heinous crimes. Popular myths held that the drug made black men homicidal as well as exceptional marksmen. Perhaps the most outrageous claim was that the drug rendered this group unaffected by .32-caliber bullets. Incredibly, these ridiculous assertions were actually believed. They prompted some southern police forces to switch to a larger .38-caliber weapon in order to deal with the mythical black, cocainized superhuman.[18]

As concerns grew about the purported widespread drug use by despised groups, several states passed laws restricting access to opioids and cocaine, making them available only by prescription. Simply put, whites with means or access to a physician still could get their drugs of choice without running afoul of the law, but others no longer had such rights.

The federal government got involved, too, which was unheard of at the time. In 1914, Congress debated whether to pass the Harrison Narcotics Tax Act, one of the country's first forays into national drug legislation, which sought to tax and regulate the production, importation, and distribution of opium and coca products. Proponents of the law saw it as a strategy to improve strained trade relations with China by demonstrating a commitment to controlling the opium trade. Opponents, mostly from southern states, viewed it as an intrusion into states' rights. They had prevented passage of previous versions.

Now, however, the law's proponents had found an important scapegoat in their quest to get it passed: the mythical "Negro cocaine fiend," which prominent newspapers, physicians, and politicians readily exploited. At congressional hearings, "experts" testified that "most of the attacks upon white women of the South are the direct result of a cocaine-crazed Negro brain." It worked. When the Harrison Act became law, proponents could thank the South's fear of blacks for easing its passage.

It's important to point out that the Harrison Act, like most state laws, did not explicitly prohibit the use of opioids or cocaine. These drugs remained readily available to those with social capital. For others, enforcement of the new law quickly became increasingly punitive. The law helped set the stage for passage of the Eighteenth Amendment, prohibiting alcohol, in 1919 and, ultimately, for all our drug policy moving forward. As important, the racial rhetoric that laced those early conversations about drug use didn't just evaporate; it endured and evolved, reinventing itself decade after decade, from reefer madness to crack babies.

OPIOIDS IN AMERICA TODAY

I am concerned that today's sensationalistic media coverage of the opioid crisis continues a long, awful tradition of exploiting ignorance and fear to vilify certain members of our society. In the process, civil liberties become collateral casualties as new, even more restrictive drug laws are passed.

As with previous "drug crises," the opioid problem is not *really* about opioids. It's mainly about cultural, social, and environmental factors such as racism, draconian drug laws, and diverting attention away from the real causes of crime and suffering. As you'll discover throughout this book, there's nothing terribly unique about the pharmacology of opioids that makes these drugs particularly dangerous or addictive. People have safely consumed them for centuries. And, trust me, people will continue to do so, long after the media's faddish focus has faded, because these chemicals work.

Fatal overdose is a real risk, but the odds of this occurring have been overstated. It is certainly possible to die after taking too much of a single opioid drug, but such deaths account for only about a quarter of the thousands of opioid-related deaths. Contaminated opioid drugs and opioids combined with another downer (e.g., alcohol or a nerve-pain medication) cause many of these deaths.[19] People are not dying because of opioids; they are dying because of ignorance.

Also, addiction to opioids is far less common than the scare stories suggest. We have been inundated with cherry-picked accounts that portray sympathetic white opioid addicts who developed addiction through no fault of their own. In reality, less than one-third of heroin users and less than one tenth of people prescribed opioids for pain will become addicted.[20] I absolutely agree with the observation of Stephen King, the noted ex-drug user and renowned writer, that "beating heroin is child's play compared to beating your childhood."[21]

But unfortunately, despite the fact that much of the reporting on opioids is bullshit, the media coverage continues to be relentless. Even though little if any factually accurate information about drugs is presented, these attention-grabbing stories inform us that decent white folks are the real victims of this tragedy, and an "evil" drug such as heroin, fentanyl, oxycodone, or some other opioid is to be blamed. If not the drug itself, then something else—reckless doctors; degenerate, nonwhite drug pushers; or "big pharma"—is responsible for the suffering.

These messages haven't been lost on politicians, law-enforcement officials, or anyone else not living under a rock. That is one reason why public officials, especially from those states where opioids are blamed for practically everything that ails them, have pushed for increased amounts of funding for opioid-addiction treatment. Even law-enforcement agencies now advocate moving beyond the arrest-first approach and are connecting users to treatment.

But the seemingly humane approach—providing treatment to all users—is only part of the picture. Multiple states have passed legislation that enhances penalties for opioid infractions. In some states, prosecutors have begun leveling murder charges against drug dealers, friends, acquaintances, or anyone suspected of facilitating the acquisition of drugs by someone who died from an overdose.[22] At the federal level, a convicted person will receive a twenty-year mandatory-minimum prison sentence for distribution of heroin or fentanyl resulting in death or serious bodily injury.

The widespread support for this "be compassionate with some" and "get tough on others" approach never ceases to amaze me. Even popular journalist Malcolm Gladwell enthusiastically backed this course of action for dealing with the opioid situation. In a recent *New Yorker* piece, he wrote, "Manufacturers and distributors [of illegal opioids] belong in prison, and users belong in drug-treatment programs."[23] Gladwell falls right into the simple-minded trap of the "drug dealer bad and drug user good" dichotomy. I wonder if Gladwell believes anyone who drinks alcohol belongs in treatment. Does he think the occasional marijuana smoker should seek treatment as well?

In a real sense, the new "get tough on opioids" policies have been fueled by the mistaken perception that most illegal opioid dealers are black or Latino. Consider the remarks made by then Maine governor Paul LePage at a town hall forum in 2016. The governor reassured attendees that his beef was not with Mainers who merely "take drugs." Bear in mind that Maine is the whitest state in the union. His outrage, LePage said, was aimed squarely at out-of-state drug dealers: "Guys with the name D-Money, Smoothie, Shifty . . . they come from Connecticut and New York, they come up here, they sell their heroin, they go back home." But, LePage warned, before these packs of mythical drug pushers head home, they usually "impregnate a young white girl."[24]

Wow. This is America . . . in the twenty-first century. WTF.

Today, most Americans, even those who share LePage's views, are not so stupid as to state these views aloud at a public gathering. Still, LePage's comments, replete with racial paranoia and condescension toward white women, are not just sinisterly similar to the scare tactics used over a century ago; they also influence decisions on drug policy and enforcement, even now. Recent federal data back this claim: more than 80 percent of those convicted of heroin trafficking are black or Latino, even though most heroin sellers are white.[25]

The legend of heroic white public officials who vow to protect white women from drug-crazed black men is as old as the country itself. With each successive generation, the story is modified to accommodate the current reviled drug. But make no mistake about it, this legend is built on the dead and incarcerated bodies of countless black men. Baldwin once wrote, with chilling precision, "We made a legend out of a massacre."[26]

RETURNING TO OUR classroom discussion . . . I reminded my students that whenever the public is gripped with fear, even if it's trumped-up, the government responds by infringing upon fundamental liberties. Think about 9/11 and the Patriot Act. Of course, *necessity* is the excuse given for each infringement; it doesn't matter whether it's restricting free speech, banning gun ownership, or prohibiting heroin use. And let's not deceive ourselves, there's a whole lot of money made on publicizing the crisis that precipitated public fear as well as on enforcing the subsequent restrictions. Stories about the opioid crisis sell everything from newspapers to documentary films, and without drug-law violators to punish, a tremendous number of people would be unemployed.

If there isn't vigorous and continuous resistance to governmental intrusions on freedom, the rights guaranteed by our noble founding documents will be steadily eroded. I reminded my students that it is their responsibility to fight each day for these rights. If they don't, we will lose them.

The corrections officer's abrupt, loud knock on the thick glass window signaled the end of our class period. My students stoically prepared to go back to being inmates. And I prepared to leave, as I did every Friday night, with the same sinking feeling that our ware-housing of millions of Americans behind bars is simply vicious.

On my fifteen-minute silent walk from the prison to the Ossin-ing Metro North Train Station, I pondered the same questions I did the previous week. Is it not cruel to teach my incarcerated students lofty ideals, knowing as I do that such ideals don't apply at their cur-rent address, especially if this is the *final* address? Is it not cruel to link these students to a world that's not available to them on the inside or outside? Am I not merely an accomplice to our cold-hearted system of justice that preys on the poor and the inconvenient? Am I just another liberal checking the "community-service box" to feel better about myself, like so many of the other volunteers I've come across?

On the train ride home, the mood was festive due to the many young, alcohol-intoxicated passengers headed into the city for a night of partying. Some were so kind as to offer me a drink. "That's mighty American of you," I said, "but I must decline because it's not my drug of choice." But equally important, my mind was still trou-bled, ruminating on my students and our discussion. As usual, these distressing thoughts stayed with me for several days, disrupting my sleep and depressing my mood. Opioids, such as heroin, help to lessen the anguish. They also come with the added benefit of pro-ducing blissfulness. It angers me that I can't enjoy these opioids as freely as my fellow passengers enjoyed their alcohol. Prohibiting one's pursuit of pleasure for baseless reasons is wrong and decidedly un-American.

2

Get Out of the Closet:
Stop Behaving Like Children

Our lives begin to end the day we remain silent
about things that matter.

Martin Luther King Jr.

I dreaded getting a colonoscopy. Sitting at the edge of a narrow hospital bed, in that antiseptic room, I was stressed the fuck out. To make matters worse, they had me change into a flimsy gown designed to make a fifty-year-old feel like a child.

It was July 2017; I was starting my second year as department chair and was feeling about as comfortable in that role as I was about having this procedure.

"Are you allergic to any medications?"

"Nah," I answered casually, trying to conceal my nervousness.

"Do you have diabetes? High blood pressure?" This young brown-skinned nurse continued reading aloud in a rapid-fire manner from a checklist. "Are you currently taking any medications? Do you drink alcohol? You smoke?"

My string of "nope" responses was interrupted with a question. "Hold up, smoke what? Certainly not tobacco."

"How about anything else?" she asked.

"Yeah, I smoke a little weed every now and then," I replied, which didn't seem to surprise her. Perhaps because I'm a dreadlock, and some people believe that we all smoke weed. Or perhaps because most states now allow patients to use cannabis for specific medical conditions; New York, my home, is one such state. Or perhaps because a growing number of states, New York not included, now allow adults to purchase and use the substance for recreational purposes. In other words, these legal developments have certainly made it less risky, and even fashionable, for some adults to be out of the closet about their cannabis use.

Still, at that moment, I was faced with a decision about whether I should fully disclose the extent of my drug use to this medical professional. I knew that knowledge of my drug use wouldn't impact whether the colonoscopy would take place. That was a foregone conclusion; it was going down within minutes. I also knew that disclosing the full extent of my drug use was risky because doing so could trigger an intensive and intrusive investigation by the Administration for Children's Services (ACS) for "parental neglect," a catchall term that can cover anything from poverty to drug use. At the time, Malakai, my youngest child, was sixteen years old. ACS wouldn't have cared that he was well cared for, nor would they have cared about his being healthy, relatively happy, and doing exceedingly well in school.

Since 2009, I have served as an expert witness in multiple court cases during which ACS sought to remove youngsters from their mothers' custody simply because the mothers tested positive for marijuana on a urine toxicology exam and/or acknowledged prior marijuana use. Mind you, marijuana use in and of itself—or any other substance use, for that matter—does not compromise an individual's ability to perform parental duties. And certainly, a drug-positive urine test does not provide any information about the user's current state of intoxication or her or his ability to function appropriately. It's like saying, "I saw an empty bottle of beer in your house, therefore you're an unfit parent." It's ridiculous.

Many thoughts raced through my head as I considered how much to divulge about my personal use. Thoughts like, "Shit, I could be putting myself and my child at risk, not from drug effects but from the disruptive presence of ACS." At this point in my life, I had seen too many families pushed to the brink of disintegration by heartless bureaucrats. And once ACS is in your life, it's extremely difficult to get rid of them. They can be like a bad relationship that won't end, no matter how many times you change your number or block the person on Facebook, Twitter, or Instagram.

In one exemplary case, a black mother of five had each of her children temporarily placed in the custody of a relative after she and her newborn were found to have marijuana by-products in their systems. The infant was not born prematurely. She wasn't underweight. She wasn't experiencing withdrawal symptoms. She didn't require a specialized level of care as a result of her mother's drug use. In fact, the evidence indicated that all the children were thriving in their mother's care. Despite these facts, ACS filed petitions alleging parental neglect, an action which could have resulted in the permanent removal of the children from their mother's custody if the judge had agreed.

As the legal jockeying played out over the next *two years*, with the stress of the case hanging over her head, the mother watched her infant achieve developmental milestones. ACS conducted multiple unannounced visits, during which they assessed the home and conducted individual interviews with the older children. The children's ages ranged from newborn to sixteen years. Can you imagine being told that your child will be "interviewed" outside your presence by some wet-behind-the-ears twenty-five-year-old social worker? ACS sought to turn the children into informants, probing these young people about whether they had previously observed their mother smoking marijuana. They hadn't. There was no evidence establishing impairment or imminent risk to the newborn or any of the other children.

And thankfully, a thoughtful family-court judge weighed all the evidence and dismissed the petitions of ACS. The family was reunited after this long legal nightmare. I was relieved and heartbroken for them at the same time.

Incidents such as this one filled my head with trepidation as I considered how much to reveal about my own drug use. At the same time, though, I was tired, tired of being dishonest. I was tired of pretending that marijuana was different, from a biological perspective, from something like heroin. Why is it OK for me to admit having used marijuana but not heroin? I know why, of course. Because most people have been led to believe that heroin is inherently a dangerous drug, whereas pot is just a harmless giggle. It's frustrating. Consider the remarks made by senator Bernie Sanders at the start of 2018 on this subject: "Marijuana is not the same as heroin. No one who has seriously studied the issue believes that marijuana should be classified as a Schedule I drug beside killer drugs like heroin."[1]

As politicians go, Sanders seems to be a fair-minded, well-intentioned individual. But his drug perspective is ignorant. Here's why: In order for a drug, any drug, to produce an effect in the brain, it must first bind to a unique site recognized by that drug. This site, this "receptor"—a specialized structure that recognizes and responds to a particular chemical—is endogenous, meaning that it's in all of us. We also have an endogenous chemical that binds to each of these receptors. That means, each of our brains contains heroin-like and THC-like chemicals and their corresponding receptors.

Why, you may ask, would our brains contain a heroin-like substance? Or even a marijuana-like substance? Well, heroin belongs to a class of chemicals called opioids, and opioids participate in myriad important biological functions. For example, they relieve pain, reduce diarrhea, and induce sleep. It's not difficult to see the life-sustaining value of this class of chemicals. Similarly, the marijuana-like chemical in the brain plays an important role in food intake and coordinating bodily movements, as well as other vital functions.

These chemicals, or more precisely, their endogenous relatives, are critical for our survival. Still, neither heroin nor marijuana is inherently more evil than the other. It's true that heroin, for example, will more readily cause respiratory depression than marijuana will. It would be a mistake, however, to conclude that heroin is more evil than marijuana is. Smoking marijuana is far more likely to cause temporary paranoia or disturbing perceptual alterations than heroin administered by any route is. If someone were suffering from dysentery, a condition that remains a major cause of death in countries that have inadequate health resources, heroin would be the obvious lifesaving choice. The point is that all drugs can produce both negative and positive effects. So to act as if marijuana is intrinsically or morally superior to heroin—or any other drug, for that matter—highlights the ignorance of the holder of this belief. Such ignorance also decreases the odds of people honestly reporting the use of drugs other than marijuana because of the stigma attached to so-called harder drugs, such as heroin.

For at least the past five years, I've felt a nagging sense of guilt about complying with the requirement to lie about my current drug use. This requirement, this falsehood, is adhered to particularly by people, such as me, whose livelihoods lie outside the arts. Sure, it's now acceptable—and even saluted in some circles—for a person to disclose honestly *past* drug use. (Even presidential candidates can now acknowledge once having used an illegal substance during their youth.) But this same person would be severely criticized if he admitted to snorting cocaine with his wife during a recent vacation in Portugal.

Why should this person be condemned or looked down upon? Because cocaine is evil? Well, we know this isn't true for any drug. We also know that cocaine was the first local anesthetic discovered and that without it we might not have drugs such as lidocaine, another local anesthetic, widely used on the skin to relieve pain and itching. Lidocaine is also used to dampen dental pain. I couldn't imagine having dental work done without the aid of a local anesthetic. Big ups to pharmacology! Even cocaine itself is still used to this day in medicine for some mouth, nose, and throat procedures.

But it is true that cocaine is banned for recreational use in the United States. So perhaps it's reasonable to rebuff a politician, or anyone, who confesses to recent recreational cocaine use. After all, the person did break the law, right? Well, it's not that simple. First, the vast majority of us are lawbreakers. Who hasn't exceeded the speed limit at least once when driving an automobile? I certainly have. And if you've had premarital sex in the states of Idaho, Mississippi, North Carolina, or Virginia, then you're a criminal, because fornication remains against the law in those states. Second, some laws are unjust, and conscientious people deliberately disobey them to draw attention to that injustice. On December 1, 1955, Rosa Parks famously refused to relinquish her seat to a white passenger as the law dictated. Parks's act of civil disobedience in defiance of an unjust law has not only made her a cultural icon but also contributed to the country becoming a more perfect union. Finally, say the cocaine use in question took place in Portugal, where all drugs are decriminalized. That's right, *all* drugs, including, cocaine, heroin, methamphetamine, 3,4-methylenedioxymethamphetamine (MDMA, a.k.a. ecstasy), everything.

Still, decriminalization is not legalization. The two are often confused. Legalization permits the legal sale, acquisition, use, and possession of drugs. Most countries' policies regulating alcohol and tobacco, for those of the legal age, are examples of drug legalization. By comparison, the Portuguese version of decriminalization states that the *sale* of drugs is illegal and is a criminal offense. Importantly, however, under this decriminalization scheme, the acquisition, possession, and use of recreational drugs for personal use—defined as quantities up to a ten-day supply—are not criminal offenses. This means that Americans who travel to Lisbon may partake in recreational drug use without breaking criminal law.

Be that as it may, the question remains. Is it possible for Americans to disclose cocaine use, or that of any other banned drug, without being ridiculed or stigmatized, even if the drug use occurred in a decriminalized country, such as Portugal? That was the question facing me as the nurse's gaze became more piercing. Well, I thought to myself, "I've been to Portugal."

In August 2016, I went to Portugal for the first time in connection with the Boom Festival, where I served as a volunteer with an outfit called Kosmicare. Boom is a biennial gathering held in the beautiful surroundings of Idanha-a-nova Lake, about a two-and-a-half-hour drive northeast of Lisbon. There, for nine days, forty thousand psychoactive substance and music enthusiasts commune together, in Europe's answer to North America's Burning Man. But unlike Burning Man attendees, Boomers can enjoy their psychoactive substances without fear of arrest.

What's Kosmicare? The first line on their website tells you pretty much all you need to know: "a safe place for grounding galactic energies and intense experiences." Composed of about three dozen volunteers, Kosmicare primarily functions to help festivalgoers deal with "bad trips" and minimize drug-related harm. Theoretically, it's a great idea.

We, as volunteers, arrived a couple of days ahead of the festival and participated in a daylong training session. The training appropriately emphasized the fact that volunteers were not therapists. We were not there to fix people. We were there to help ensure the comfort and safety of attendees experiencing unpleasant or disconcerting drug effects.

At the end of the training day, a long, scorching one, with a peak temperature of 40 degrees Celsius (104 degrees Fahrenheit), we gathered in a circle. Led by a couple in their mid-to-late fifties who had clearly done this on numerous previous occasions, we were instructed to close our eyes as they walked past us shaking a large bundle of burning sage and chanting incoherently. We were then instructed to pair up, gaze into the eyes of our partners, and tell them that they were seen, I mean *really* seen. Tell them that they were beautiful, *really* beautiful; and tell them they were Love, *real* Love. Then, we turned to Mother Earth and told her how grateful we were for her generosity in allowing us to inhabit her.

As the séance-like ceremony drew to an end, the dominant thought in my head was, "It's nice to know that weirdos are not unique to the United States." I also quietly wondered, "What the fuck?!" and attempted to avoid eye contact with anyone. Too late. Niamh, my Irish friend, shot me a penetrating look that communicated, "This is some crazy shite, dude. If *you* don't tell anyone, *I* certainly won't. Cool?" I returned the look, and we have never spoken of it since.

All the flakiness aside, though, I was struck by how well the members of the broader Boom community treated one another. People openly shared everything from laughs to concerns to food to drugs to shelter to sex, everything. There was a strong sense of kinship. People were also permitted the freedom to be whomever they chose to be, without judgment. This part of the experience was truly beautiful; it was inspiring. I came away wanting to be a better person, wondering how I might incorporate what I had experienced in my daily life.

The Boom organizers also provided free, anonymous drug-purity testing services, another indicator of how serious they were about keeping attendees as safe as possible. Attendees could submit a small portion of their drug for testing in order to know the composition and purity of the sample. In this way, they would know whether their drug contained impurities or adulterants, which are usually more dangerous than the drug itself. This service is potentially life-saving.

Unfortunately, I had little free time to enjoy all that Boom had to offer because of the demands of Kosmicare. Young people experiencing unpleasant drug effects, including anxiety, paranoia, and insomnia, trickled in and out of our service. Some had taken too much of a particular substance or too much of multiple substances. Others were simply in need of a sympathetic ear or a calm place to lay their head. The following is a typical case.

• • •

AT 4:15 A.M., PAULO, a twenty-five-year-old Portuguese man, came
to Kosmicare because he couldn't sleep. Paulo had taken an MDMA
tablet—dose unknown—at about 10:30 p.m. the previous evening;
an hour later, he had ingested two doses of LSD, again dose unknown. During the course of the assessment, it became clear that
Paulo was not distressed about the drug effects, but instead, he worried tremendously about his girlfriend's harsh judgment regarding
his "irresponsible drug use." The girlfriend was due to arrive at Boom
later that day, around 1 p.m.

Paulo was lucid, appropriate, and funny. He even joked that his
girlfriend could benefit from the counseling services of Kosmicare.
After a two-hour nap and a few cigarettes, he expressed slight embarrassment at having sought our services and left to rest in his own tent.

Thankfully, fewer than 1 percent of Boomers needed our services. This is in line with the preponderance of evidence that most
drug use occurs without problems. Even so, at least one person died
from what was said to be cardiac arrest, possibly brought on by
heavy use of multiple drugs and extreme heat. It's not clear to me
exactly which drugs were involved and to what extent they may
have contributed to the person's death. Nonetheless, this tragedy
underscored for me the immense need to better educate the public
about how to use drugs in a manner that enhances desired effects
and minimizes adverse outcomes.

By the end of my Boom experience, I was mentally and physically
exhausted. I was also starving and had lost nearly ten pounds. So I
rationalized that I deserved a nice meal and a fancy Lisbon hotel
room on my final night in Portugal. As I unpacked shortly before

dinner, I discovered among my belongings a 10 mg tablet containing 2,5-dimethoxy-4-bromophenethylamine (2C-B). A friend whose research focuses on this compound had given me a few of these psychedelic pills at Boom to try. I did. They were enjoyable. I was excited to find this remaining pill. I couldn't take it back with me to the United States, where its possession is a criminal offense, so I would have to ingest it that evening or discard it. The latter choice seemed sacrilegious.

I showered, took the pill, and went to meet friends for dinner in the center of Lisbon. It was a sweet, warm evening, accentuated by the sea of friendly faces that passed as we dined outdoors on a side street. It had been about thirty minutes since I had taken the 2C-B. I felt a slight euphoric buzz. It was nice.

We laughed and reminisced about our time in the wild. I was *so* relieved to be back in civilization and was *so* looking forward to lying on the big comfortable hotel bed. I was also extremely grateful for my friendship with my dinner companions. They were truly generous spirits, people who had donated their time and effort for two weeks in order to help recreational drug users partake in a safer, more comfortable, nonjudgmental environment.

An hour had passed since I had taken the 2C-B. My appetite was waning, but my admiration for my friends was intensifying. The effects of 2C-B were really kicking in.

After dinner, we returned to the hotel, and I prepared for bed. Still mildly under the influence of 2-CB, I reflected on the evening and my time at Boom. In the background, Nina Simone was singing her soul-wrenching song "Why?," which laments the murder of Martin Luther King Jr. Something about the combination of 2-CB–related effects and Nina's heartfelt voice made me *feel* the song as if for the very first time, even though I had listened to it since I was a child. I was able to make connections between the love and freedom King advocated for everyone and the love and freedom we all experienced at Boom: it came down to the protection of personal liberty and the unalienable right to pursue happiness. The Portuguese government has taken a step in the right direction with its drug-decriminalization policy. The U.S. government, by contrast, has yet to stand up to the promise of the Declaration of Independence. Nina asked earnestly, "Will my country fall, stand or fall? Is it too late for us all?"

These questions dogged me when I returned home. I had to decide whether I'd continue to participate in the dishonest endeavor of concealing and denying my current drug use. I knew all too well that the fear of being exposed as a drug user forces many responsible adults straight into the closet. Perhaps even worse, this fear allows the broader society to retain its caricatured view of drug users as irresponsible and troubled souls, who then can be summarily demonized and marginalized. What kind of man would I be if I didn't stand up on behalf of these individuals? They are my people; they are our people. What kind of man would I be if I didn't stand up for liberty for all?

LIBERTY MEANS RESPONSIBILITY

Liberty—and, by extension, freedom—is virtually impossible without responsibility. I am a responsible person by most traditional measures. As a responsible person, I must be cognizant of the consequences of my actions on people and the environment. If my actions hinder others from exercising their freedoms, then it is incumbent upon me to alter my behavior. Responsible people make such adjustments. But being responsible isn't easy. Responsibility requires a considerable amount of self-inspection and a healthy sense of respect for fellow humans. It takes a tremendous amount of ongoing work. Grown-ups put in the work because it affords us liberty; it entitles me to live my life according to my own values.

Another reason we do the work is that the alternative is simply too revolting: being told by the government what to think, what to put in our bodies, and how to live. I recognize that many have accepted this bargain because they may not want to think for themselves. Others may be frightened by some perceived or real enemy. In their minds, abdicating responsibility to the government will keep them safe. "The average man does not want to be free. He simply wants to be safe," H. L. Mencken noted. Of course, this is an illusion. There is no such thing as being completely safe. Any life worth living is not without risk. So I would suggest you live your life as you see fit. At least you will be living.

In the United States, liberty is such an important right that it is protected by the Constitution. The liberties enshrined in this document ensure that I am free to eat sugar-laden shortbread cookies, despite my family history of diabetes. A high-sugar diet can raise my risk of dying from a diabetes-related illness. Taking this into consideration, it behooves me, a responsible adult, to weigh the risk-to-benefit ratio when deciding whether or not to have that next cookie. But there is a role that the government should play. It should provide accurate, nonbiased information that sheds light on the potential risks and benefits associated with a product. Warning labels on tobacco and alcoholic products, ingredients lists, and nutrient-breakdown labels on food products are just a few examples.

Each and every day, we all are faced with potential risks and must make risk-to-benefit calculations repeatedly. This is a basic fact of life. Our right to make decisions based on the outcome of these calculations is not outlawed by the government, *except* when it comes to certain recreational drugs.

As a scientist, I find this exception particularly frustrating, even hypocritical. The justification for restricting specific drugs is often related to the purported inherent dangers posed by these chemicals. Heroin use, for example, is said to be inherently more dangerous than other legal activities such as gun or car use are. Really? Guns, let's not forget, are specifically designed to kill. This is not to say that every owner purchases a gun with this goal in mind. As a budding gun hobbyist, I know that's not true. Still, each year there are about forty thousand gun-related deaths, and more than half are suicides.[2] In 2017, heroin-involved deaths reached an all-time peak at just over fifteen thousand, a number well below that of gun deaths.[3] (Again, it's important to note that most of these heroin deaths occurred because the drug was contaminated with a far more potent fentanyl analog or because it was combined with another sedating drug, such as alcohol or sleeping pills.) I am not arguing that guns should be banned. Rather, I absolutely agree with journalist and anti-lynching activist Ida B. Wells when she exhorted, "a Winchester rifle should have a place of honor in every black home, and it should be used for that protection which the law refuses to give." Thankfully, the Second Amendment protects this right. My point is that no sane person can argue with a straight face that heroin use is inherently more dangerous than gun use. At the very least, this should raise the question, Why is it that guns can be legally purchased but heroin cannot?

Yeah, I know, some of my liberal friends are not feeling the above example and would be just as happy to ban both guns and drugs. So consider an example involving automobiles. Driving a car, and even riding as a passenger, is a potentially deadly activity. In 2018, more than forty thousand Americans lost their lives in car accidents, a number that has remained relatively stable for the past three decades.[4] But no one is calling for a ban on automobiles; nor should they. The idea is absurd, as is the banning of recreational drugs.

Even when heroin and alcohol use are compared, it's difficult to make a compelling case for the unique dangers posed by heroin. Both of these drugs can cause respiratory depression and death if large doses are administered to naive or infrequent users. And both heroin and alcohol can lead to unpleasant withdrawal symptoms when chronic use is abruptly discontinued. More important, though, one can die from alcohol withdrawal but not from heroin withdrawal. In addition, of the two drugs, only alcohol can cause severe liver damage leading to death. Each year in the United States, there are nearly one hundred thousand alcohol-related deaths.[5] Again, I am not making an argument for restricting the availability of alcohol. Trust me, I cherish my right to consume this anxiety-relieving beverage, even though I rarely consume it. Without periodic alcohol consumption, it would've been hellish to get through the countless receptions I was required to attend as department chair. Furthermore, there are multiple beneficial effects associated with responsible alcohol use.

Still, the above raises the question, Why is alcohol legal, whereas heroin is banned? Again, there are far fewer heroin-related deaths than there are those associated with alcohol. This makes it difficult to justify a ban on heroin based solely on public health concerns. Some speculate that heroin deaths would surpass the number for alcohol were heroin legal, but the available evidence contradicts this. In Portugal, for example, the number of heroin users actually dropped from 100,000 to 25,000 after that country decriminalized all drugs. Portugal has the lowest rate of drug-induced death in

Western Europe. To put that in numeric terms, in 2016, 60 people in Portugal died, which translates to 6 deaths per million. This number is a small fraction of the American death toll, which was 312 deaths per million.[6]

What is also clear is that as long as heroin and other drugs are banned, users will be less willing to disclose information about their use even to medical professionals. Imagine your doctor's response if you told her that you used heroin last weekend. This silence leads to more unnecessary health risks because it contributes to individuals being more susceptible to misinformation about heroin and compromises the patient-provider relationship.

When I became department chair, a colleague informed me that professor Robert Bush, a well-liked previous chair in our department, had died from what was characterized as a narcotics overdose in 1972.[7] People speculated that heroin was the causal agent. Bush was fifty-one, my age as I write this. He also belonged to a marginalized group: I'm black; he was gay. He taught a course on drugs; I now teach such a class. He resigned his chairmanship after a dispute with the administration but resumed the role shortly thereafter; I did, too. These parallels increased my curiosity about the circumstances surrounding Bush's death.

I wondered about the extent of his opioid knowledge. I wondered if ignorance had played a role in his death. I know for certain that lack of information plays a huge role in the annual number of heroin-related deaths. Many users don't even know if they actually have heroin because some dealers pass off other potentially more dangerous substances as the drug. Also, some people unwittingly combine heroin with a benzodiazepine such as Xanax—or another sedative— and thus increase the risk of overdose, especially in nontolerant individuals. Just as importantly, street-heroin potency varies widely, and it varied even more in the 1970s. This variability adds even more uncertainty about the strength of effects the user could expect. I wondered if any of these factors contributed to Bush's death. Whatever the case, I am certain that if heroin were legal, more people would be informed about a wider range of lifesaving information. The current legal restrictions impede communication between users and health-care professionals, as well as communication between more knowledgeable heroin consumers and the general public.

BACK IN THE colonoscopy exam room, the nurse continued to jot down my responses to her questions. "I also, from time to time, do a little cocaine, heroin, and molly," I told her. "Huh?!" she exclaimed, sudden interest in her voice. Her wide-open eyes stared at me with disbelief and without embarrassment. I wanted to laugh but didn't. Instead, I put her at ease. I reassured her that I was not the stereotypical drug addict she had most likely learned about in her miseducation. "It's cool," I said. "I haven't taken anything recently." I also described my profession and explained that I have published extensively on the topic of drugs. Still looking uncertain, but trying to appear cool, she meekly replied, "Oh, OK." Autopilot now turned off, she asked about where I worked, my travels, my interests. She now saw me.

In the end, I knew I had done the right thing by coming out of the closet. Sure, it's risky, even frightening. But people like Rosa Parks—and so many others—faced far more dangers in their efforts to rid our society of unjust laws. Knowing that I had benefited directly from their gallant efforts, my decision was simple. Remaining in the closet about my drug use felt cowardly, dishonorable. Why should I be required to conceal an activity that I enjoy, especially if it doesn't negatively impact others? I am not a child, nor will I be treated as such. Living in American society, where black men are too often relegated to childlike status due to American racism, it's too damn hard to stomach being made to feel like a child in yet another domain. Thus, out of the closet I will forever remain.

3

Beyond the Harms
of Harm Reduction

The road to hell is paved with good intentions.

English Proverb

Jesus loves every 1 of U—read the banner pulled by the airplane that circled above us on a dreary Sunday afternoon as I stood in the middle of Manchester's Heaton Park taking in the sights and sounds of the 2018 Parklife Festival. Parklife, like so many other festivals, takes place over several summer days during which mostly well-to-do young people commune, listen to music, dance, and generally enjoy themselves. Some even take recreational drugs in hope of enhancing the experience.

Though I'm not particularly religious now, during my youth, I spent many Sundays attending a Southern Baptist church in Florida. I was thoroughly indoctrinated with the belief that Jesus took a particular interest in looking after the poor, the least among us, and I was taught to do the same. The airplane banner reminded me of this responsibility. I wondered if its sponsor would endorse the idea that looking after each other extended to using our knowledge, skills, and platforms to help keep people safe and healthy, even if they use drugs.

I had been invited to attend Parklife by a UK organization—The Loop—that was attempting, in its own secular way, to live up to these ideals. The Loop provides harm-reduction services to people who use drugs. Simply put, *harm reduction* strategies seek to reduce negative consequences associated with drug use. Providing clean needles and syringes to an intravenous heroin user is an example of harm reduction because this decreases that person's chances of contracting a blood-borne infection from sharing contaminated injection equipment. Instructing a festivalgoer to drink lots of fluids, to stay well hydrated, if the individual takes a diuretic drug such as MDMA or methamphetamine is another example of harm reduction.

Outside the drug world, each one of us, on a daily basis, takes measures to prevent illnesses and to improve our health and safety. We brush our teeth, wear seatbelts, use condoms, exercise. We don't call it harm reduction; we call it common sense, prevention, education, or some other neutral name. The point is that the term *harm reduction* is used almost exclusively in connection with drug use and has negative connotations. It often conjures up an image of an intoxicated and belligerent individual imprisoned by the use of a substance or some other unfavorable image depicting a drug user in need of being saved. In addition, the term implies that services (e.g., education) geared toward drug users should focus primarily on the potential harms caused by drug use and on strategies to reduce them. In a nutshell, the term *harm reduction* obfuscates the fact that most people use drugs to enhance experiences, to bring about euphoria—for pleasure.

"HARM REDUCTION" HAS TO GO

I walked around the park grounds checking out the jubilant crowd of young partygoers. A gospel-sounding, organ-dominated groove played in the background. It was a familiar sound. I couldn't name it; I also couldn't ignore it. My attention, though, was being beckoned by one of the volunteers from The Loop. Excited, this kind bloke told me about the harm-reduction services they were providing at the festival. His huge handlebar moustache and crew cut evoked thoughts of Confederate soldiers, flags, monuments, the whole nine. For a southern black person, this wasn't a good first impression. But I knew this was *my* issue, not his, so I did my best to focus on anything other than his facial hair. I checked out his gear. It was casual—red-rimmed bifocals, a tie-dyed T-shirt, and cargo shorts—and seemed incongruent with his hairstyle. In an attempt to reciprocate the kindness and respect he showed me, I tried to listen more attentively.

But it was difficult because a recording of Al Green's 1972 classic, "Love and Happiness," blasted from the sound system. "Something that can make you do wrong, make you do right," Green sang with a great deal of sorrow combined with the rapture expressed by a true believer. His brilliant treatment of the anguish and joy that can result from being in love hit me like an amphetamine.

I reflected on the idea of harm reduction. It doesn't capture the complexity associated with grown-up activities such as love or war or drug use. Instead, it preoccupies us with drug-related harms. And the connection between harms and drug use is reinforced repeatedly through our speech. This connection in turn narrows our associations, conversations, feelings, memories, and perceptions about drugs and those who partake. Perhaps even worse, it relegates drug users to an inferior status. Surely, only a feebleminded soul would engage in an activity that always produces harmful outcomes, as the term implies.

In that moment, listening to Al Green testify, alongside my handlebar-moustached host, I felt certain that the term *harm reduction* had to go. It had worn out its welcome. We needed a new term, new language; because the language we use shapes how we think and behave. We need to think about drugs and behave in a more nuanced manner. We need to cut the bullshit and stop pretending drugs inevitably—and only—lead to undesired outcomes.

I pondered the question of what term or phrase I would use as an alternative. I had no idea. But I knew the updated expression had to be multifaceted. It had to be flexible enough to accommodate myriad drug effects, whether they were good, bad, or indifferent. And, like the song "Love and Happiness," it had to capture complex and even conflicting constructs. "Health and happiness" popped into my head. I liked it. It sounded like "love and happiness" but included the important word *health*, so it could be broadly applied to other activities in which we engage.

For example, traveling via car presents potential risks to one's health as well as potential benefits that impact one's happiness. Wearing a seatbelt, replacing tires so they are not worn, and making sure the brakes and windshield wipers function properly—all can be conceptualized as "health and happiness" strategies. Similarly, setting aside at least an eight-hour sleep period following a bout of heavy stimulant use is also a "health and happiness" strategy.

Furthermore, the phrase *health and happiness* reminded me of the noble ideals laid out in our Declaration of Independence. The signers unequivocally declared that it is our *unalienable right* to pursue *life, liberty, and happiness.* The bottom line is this: millions of Americans, including me, have discovered that certain drugs facilitate our ability to achieve this goal, even if only temporarily.

I recognize that I have no authority to coin a phrase for an entire field, especially one comprising many experts who have been doing this work long before I even knew the field existed. That isn't my goal. Frankly, I think there need not be a specific term for harm reduction. We already have such terms: *common sense, prevention, education,* and the like. I don't much care which term is used, just as long as it doesn't box drug use into an exclusively harmful category and it recognizes the positive features of the experience.

OPIOID CRISIS: A CRISIS OF DATA COLLECTION AND REPORTING

Unfortunately, simply replacing the term *harm reduction* won't do much to combat sensationalist media headlines that too often give the impression that death is the only outcome associated with drug use. Panic-stricken coverage of the so-called opioid crisis is an acute example. OPIOIDS RESPONSIBLE FOR TWO-THIRDS OF GLOBAL DRUG DEATHS IN 2017: UN was the title of a typical article on the subject.[1] In the piece, the author concluded that opioids were "responsible for two-thirds of all drug deaths worldwide." Really? I doubt it. I am not suggesting that fatal drug overdoses don't occur; they do. Nor am I suggesting that we, as a society, should not be concerned about such cases; we should. My point is that the evidence for this claim is weak at best. Events leading to drug-related deaths are often far more ambiguous and complex than media reports would have you believe.

In the United States, the Centers for Disease Control and Prevention (CDC) collects mortality data from death certificates, which contain the cause of death. These certificates are filled out by thousands of different people across the country. Each state determines its own standards and requirements for individuals conducting death investigations. As a result, death investigators vary widely in their training and experience. Some are medical examiners and others are coroners. Medical examiners are physicians with specialty training in forensic pathology, whereas coroners are not required to have any medical training (except in Arkansas, Kansas, Minnesota, and Ohio). Usually medical examiners are appointed by a chief medical officer; coroners are elected by the voting public. Remarkably, any eligible voter can become a coroner, regardless of their knowledge—or lack thereof—on issues related to death investigation. What's even more absurd is that most regions in the United States rely on coroners (see Appendix 1, for a detailed list by state). As you might imagine, these different standards can and do produce considerable variations in the collection and reporting of cause-of-death data, including drug overdose.

Added to this glaring systemic defect is the variety of circumstances surrounding drug-related deaths. In most cases, more than one substance is found in the body of the deceased, and the concentrations of these drugs are often not determined. Hence, it is difficult, if not impossible, to attribute the death to a single drug because we can't know which, if any, of the drugs reached a blood level that would be fatal by itself. Whenever officials or reporters claim that a particular drug caused the death, you should ask about the drug's concentrations in the body and whether other drugs were involved. A related point regarding fatal overdoses that involve multiple drugs is that individual deaths can be counted more than once. If the deceased's body contains three drugs, for example, then this can be recorded as three separate overdoses. Obviously, this "alternative" math overstates the overdose death toll and hinders the public's ability to obtain an accurate picture of the problem.

A recent report published in the journal *Science* revealed even more problems with the current system of tracking fatal overdoses.[2] In about a quarter of these deaths, no drug is listed on the death certificate. One reason for this is that some jurisdictions do not test, or test consistently, for the presence of drugs in the body of deceased individuals. Some jurisdictions may test for the presence of a select group of drugs, but not others. Also, death investigators' decisions to test for certain drugs or to attribute the cause of death to a specific class of drug are likely influenced by news reports of the *drug du jour* and by subjective beliefs about which drugs are most dangerous. These problems can be compounded by the fact that death investigators may not be trained to make determinations about the intent of a drug-related death. It's not always easy to tell whether the death was accidental or intentional, such as suicide.

Finally, one of the most important limitations of many recent analyses and reports on drug overdose deaths is that the co-occurrence of alcohol is ignored completely.[3] Large doses of alcohol alone can cause fatal respiratory depression, and when combined with opioids or other sedatives, considerably smaller doses can become lethal.[4] So, the next time you see a graph or report warning about the dramatic increase in opioid-related deaths, remember that a sizeable proportion of the purported number of deaths may be due to changes in detecting and reporting, rather than opioids themselves.

THE FENTANYL SCARE

The flaws in the current system of tracking overdoses haven't received much media attention. That's one reason it has been relatively easy to frighten and mislead the public about the current so-called opioid overdose epidemic. The recent focus blames fentanyl for many of these deaths. A stream of media headlines blare: MOTHER SHOCKED AS TASK FORCE RECOVERS ENOUGH FENTANYL TO KILL 32,000 PEOPLE,[5] FENTANYL-RELATED DEATHS CONTINUE "STAGGERING" RISE IN MARYLAND,[6] and FENTANYL OVERDOSE DEATHS IN THE U.S. HAVE BEEN DOUBLING EVERY YEAR.[7]

Let us not forget that fentanyl is a medication approved by the Food and Drug Administration (FDA) for use in treating severe pain, such as that caused by cancer. In the United States, the drug has been used for this purpose since 1960. Fentanyl is a safe and effective medication when used as prescribed. So why the sudden panic about a medication that has been on the market for nearly sixty years?

Because fentanyl is an opioid, it can produce an opioid-like high, hence its attractiveness to some recreational users. Indeed, many recreational opioid users intentionally purchase fentanyl and have happily used it for years. Philippine president Rodrigo Duterte openly admitted using the drug to get high. He has also provided ringing endorsements attesting to its pleasurable and anxiety-suppressing effects: "You feel that you are on cloud nine, as if everything is okay with the world, nothing to worry about."[8]

On the other hand, fentanyl and its analogs, such as carfentanil, are considerably more potent than most opioids, including heroin, so less of it is required to produce an effect. As long as the user is aware of this fact, this is a beneficial feature of the drug. Smaller amounts of fentanyl are needed to produce desired effects, so the drug can be more easily hidden on one's person. It's equally important to emphasize here that fentanyls produce respiratory depression and fatal overdose far more readily than heroin.

The differential potencies between these opioids has become a major topic of concern because fentanyls are increasingly sold as heroin, mixed with heroin, or pressed into counterfeit opioid pills. One of the main reasons for this is an unscrupulous practice carried out by some illicit heroin manufacturers. These individuals have discovered that they can save money and stretch their product by adding fentanyl or an analog to their heroin batches. Sadly, this information isn't always shared with low-level dealers, who sell the product to their heroin consumers. This, of course, can be problematic—even fatal—for unsuspecting heroin users who ingest too much of the substance thinking that it is heroin alone. Even so, it is important to remember that the problem isn't fentanyl per se. The problem is fentanyl-contaminated heroin and fentanyl-tainted counterfeit opioid pills. The problem is ignorance.

The Scots and Canadians are experiencing a similar issue. In both Scotland and Canada, the reported number of fentanyl-associated deaths has steadily increased in recent years. In England, there is also the constant but unaddressed worry about young people consuming contaminated party drugs. The current focus is on tablets sold as MDMA that contain an extraordinarily large amount of the substance and those that are adulterated with more potentially dangerous unknown compounds, both of which have been associated with recent deaths. In either case, drug users lack vital information about the substances they consumed. Again, the problem, readily solvable, is ignorance.

DRUG-SAFETY TESTING

One practical solution is to make drug-safety testing services widely available. It works like this: drug samples can be submitted for testing to determine their contents and the dose. This information can be given to users so they can decide whether or not to take a particular drug and how much of it to take. This sensible approach has been implemented in a handful of countries, including Austria, Colombia, Luxembourg, Portugal, Spain, and Switzerland.[9]

Bafflingly though, drug-safety testing services are not legally available in most countries, including the United States. Initially, I thought the absence of these services was simply due to health officials being unaware of the existence of this technology. I was dead wrong.

In May 2018, the Baltimore city commissioner of health, Dr. Leana Wen, and I participated in a public discussion on the "opioid crisis." Initially, I was hesitant about participating because I knew these types of forums too often morph into merely a litany of tales about the horrors of drug use. But this one seemed different. It took place before a San Francisco audience that wasn't there to hear the same overly simplistic "drugs are bad" narrative that usually dominates these discussions. I was relieved.

Also, Dr. Wen seemed caring, earnest, and trustworthy. The title Health Commissioner perfectly suited her. She recited some horrifying statistics. Between 2013 and 2017, the annual number of fentanyl-contaminated heroin deaths in Baltimore rose dramatically from just 12 to an astonishing 573, an increase of nearly 5,000 percent over four years. But her only real proposed solution was to increase the availability of the opioid-overdose antidote naloxone. Sure, this strategy would be helpful, but it alone would be insufficient and less than ideal. Naloxone is potentially useful only if another person is present because it has to be given by a bystander immediately after an overdose has occurred. Also, this strategy deals with the problem only *after* an overdose has occurred, rather than working to prevent it in the first place.

So I suggested that Dr. Wen also implement drug-safety testing services, thinking that this would be new and welcome information. To my surprise, she didn't respond. Perplexed, I engaged her privately in further discussion afterward. She was aware of this strategy, but her response was politely dismissive. This seemed really odd to me, particularly given her clear emotional concern for the health of Baltimoreans. As promised, I sent her additional information about these services and how they make drug use less dangerous.

Subsequently, on multiple occasions, I saw Dr. Wen in the media passionately reciting the same alarming statistics about the number of people dying from contaminated drugs. Not once has she mentioned drug-safety testing as a potential remedy. Thinking back on my interactions with her, I reflected on the questions she asked in private; it appeared that her beliefs were shaped by the misguided notion that any use of an illicit substance is addiction. This perspective is ignorant. It assumes all drug users need treatment. Providing naloxone is consistent with this moralistic misapprehension, but providing a service that decreases a person's chances of consuming contaminated drugs is not.

I also couldn't avoid the feeling that Dr. Wen was deeply worried that the availability of drug-safety testing services would be perceived as an endorsement of drug use. Put bluntly, she seemed far more concerned about appearing to condone use than about saving people's lives. This thought made me nauseous.

Thankfully for the people of Baltimore, Dr. Wen became the president of Planned Parenthood in September 2018. Less than a year later, she was fired.[10] I wish I could say I'm surprised.

Unfortunately, many public health officials in the United States share Dr. Wen's views on drug use. That is why drug-safety testing services are virtually nonexistent here.

A few countries have banned drug-safety testing for general use but have made an exception to this legal restriction so that testing is permitted at some festivals and nightclubs. Of course, in my view, drug-safety testing should also be available outside these settings because where these services are available, there are fewer drug-related deaths. We should care enough about the value of a human life to make these services universal, free, and anonymous. Even so, I'd be remiss not to give props to countries adopting this approach because, at the very least, it is more mature and reasonable than the inanely restrictive one taken in the United States. The partial-testing approach acknowledges what we all know—drug use is common at festivals and nightclubs—and takes a proactive step toward promoting health and happiness.

In the United Kingdom, The Loop conducts drug-safety testing under these narrow conditions. At Parklife, I saw firsthand how this works and its value. Over the course of the festival, I watched The Loop volunteers test hundreds of pills and patiently provide results, education, and safety information to medics, police, and officers from the fire brigade.

As I expected, an extremely low percentage of Parklife attendees experienced bad drug effects. This is consistent with the evidence showing that most drug use occurs without problems. Even so, because illicit drug markets are unregulated, the quality of drugs purchased can and does vary widely and sometimes leads to unfavorable reactions.

I observed one such reaction at the festival when a young woman was brought to the medical tent because she had become excessively anxious after taking a pill of what she reported to be 2C-B, a relatively mild psychedelic. The medic attending to the woman had not previously heard of the drug and was unsure of how to proceed. He brought the remaining pills to The Loop's mobile lab for testing and asked about the drug and its effects. Thanks to the advice from a senior chemist, the medic was able to establish the best course of action for his patient.

Other volunteers publicized test results through social-media postings and by placing flyers in public spaces at the festival. Every posting contained the pill's content and prominently displayed the approximate amount of the drug or drugs detected. A photo of the pill was also included, providing a clear view of its color, logo, and size.

A blue pill, pressed to look like the comic-book character Punisher, a homicidal vigilante, was quite popular at Parklife. It was sold as MDMA, and its contents didn't disappoint. It contained about 250 mg of the drug. That's a lot, more than twice the typical dose (~100 mg) needed to produce the pleasurable experience people seek from MDMA. As a result, The Loop disseminated Punisher postings that highlighted the pill's unusual strength and cautioned users to be particularly mindful about the amount ingested.

GIL SCOTT-HERON

"But I'm new here . . ." rang out of my headphones as I stood in the long line for pizza. Hundreds of carefree festivalgoers floated about. Some even checked out the Punisher pill flyer behind the counter and seemed to discuss it. I couldn't hear what was being said because the raspy, age-weathered, and worn voice in my headphones was asking, "Will you show me around?" I was listening to Gil Scott-Heron's song "I'm New Here."

Back in 1985, I discovered the brilliance of Scott-Heron when I was a young, ignorant American solider stationed in Okinawa, Japan. I bought all his albums and studied them like holy texts. Songs such as "Angola, Louisiana" described the icy brutality perpetrated by the U.S. criminal-justice system against blacks, even black children. "Johannesburg" opened my eyes to the callous cruelty carried out by apartheid South African authorities against their black population. Through his music, I learned black lives mattered long before the slogan became fashionable. Through his music, I began my process of learning how to think. He was one of the people who showed me around.

Years later, I'd learn of his drug use and the public humiliation he suffered because of it. Media reports about his use were consistently unflattering and judgmental. They never failed to mention his two convictions for cocaine possession or that time when a moralistic, grandstanding judge presented him with a pseudochoice between prison and treatment. Until his death in May 2011, Scott-Heron remained steadfast in his adamant denial of ever having had a drug problem.

In 2010, I was invited to the New York City launch party for his new album *I'm New Here* but couldn't attend because I was in San Francisco that day serving on a National Institutes of Health grant-review committee. To this day, I regret not attending that party. His management team did send me a copy of the new album. Each time I listen to the song "I'm New Here" and hear him asking, "Will you show me around?," I wish I could have used my expertise in pharmacology to give him a fraction of what he gave me.

I don't mean this in some self-important way. It's just that the song forces me to think about my responsibility regarding showing others around: to provide drug users with a few important lessons to facilitate their health and happiness. If I had to boil them down to a few tips, they would involve these four topics: *dose, route of adminis-tration, set, and setting.*[11]

DOSE

By dose, I simply mean the amount of drug taken. This is perhaps the most crucial factor in determining the effects produced by the drug. It is the *dose* of fentanyl that dictates whether a person will experience serenity or fatal respiratory depression. In general, larger doses increase the likelihood of harmful effects. This is one of the most basic principles of pharmacology.

A related issue is potency—the amount of drug required to pro-duce a particular effect. The smaller the amount needed to cause the response, the more potent the drug. I'm sure you've heard the recent claim that the marijuana available today is "ten times" more potent than the marijuana of the 1960s. The political message is that the weed of the 1960s might have been relatively harmless, but the current stuff is dangerous. Well, that's an oversimplification. If we consider only the marijuana available for smoking in the United States, potency varies widely, ranging from a low-grade product containing less than 1 percent THC to a high-grade sinsemilla containing 11 percent or more THC. The usual range of potency for marijuana seems to be 3 to 6 percent.

Equally important, the entire range of these traditional preparations has been known, and for 150 years, scientific, literary, and medical descriptions of the wide range of effects have been based on this entire range of potencies. It is true, however, that U.S. marijuana growers are becoming more sophisticated and producing more sinsemilla—that is, higher THC percentages.

The bottom line is that consumption of large amounts of cigarettes containing high-percentage THC, especially by inexperienced users, can lead to more negative drug effects, such as anxiety, paranoia, or dangerously low blood pressure. But most people, including me, tend to decrease the amount of drug inhaled when smoking products that contain high THC concentrations because less is required to produce the desired effects, in the same way that in a single sitting, you are likely to drink more beer (5 percent alcohol) than vodka (40 percent alcohol). I wouldn't say vodka is more dangerous than beer; I would simply make sure people understood the differences between the beverages in terms of dose. Dose is everything.

ROUTE OF ADMINISTRATION

Dose is not a difficult concept to grasp. Less well known is the notion that a drug must first reach the brain before it can change your mood or behavior. Obviously, we can't put a drug directly into our brains. It has to be transported through the blood to the brain. This brings me to another important basic principle: the faster a drug gets into the brain, the more immediate and intense its effects will be.

One cannot fully understand a drug's effects without considering how the drug is taken, or in the language of pharmacology, the "route of administration." Route of administration determines the speed at which the drug reaches the brain and, therefore, the immediacy and intensity of the drug's effects.

Heroin, like most drugs, can be ingested in several ways. When used in medicine to manage pain, it's often taken by mouth, usually in the form of tablets called diamorphine hydrochloride. Swallowing a drug is convenient; no special injection equipment is needed, and it can be done discreetly. This route also tends to be safer because in the case of an overdose, the stomach can be pumped, but with intranasal, smoked, or injection overdoses it can't. But a potential drawback associated with the oral route is that the high comes on more slowly and the intensity of the high might vary more than it would with other routes.

Once in the stomach, heroin is dissolved and moved to the small intestine, where it can be absorbed into the bloodstream. If you eat a large meal immediately before taking an oral dose of heroin, this will delay absorption and the onset of the drug's effects. Swallowing a heroin tablet on an empty stomach speeds absorption and produces faster effects. You've probably already experienced this phenomenon with alcohol. Most people have.

After heroin has entered the bloodstream, it must get through the liver before it can reach the brain. One of the many vital functions of the liver is to break down chemicals in an effort to make them less harmful to the body, the process known as metabolism. Our liver contains proteins that metabolize drugs, including heroin. This means that even before a dose of heroin, taken by mouth, reaches the brain, some of it will be lost due to metabolism. This phenomenon is called *first-pass* metabolism and can markedly reduce the impact of drugs taken orally.

For these reasons, some experienced drug users seeking intense highs prefer to take their drugs via routes other than orally. Snorting heroin powder, for example, bypasses the liver; blood vessels lining the nose take the drug directly to the brain. Within a few minutes, the effects of a snorted line are felt. By comparison, it could take as long as forty-five minutes before you feel the high from a heroin pill.

Injecting heroin intravenously or smoking it gets the drug to the brain even faster than snorting it does. Once injected, heroin passes through the heart and then immediately reaches the brain. The onset of the mood-altering effects is felt within seconds. But injecting heroin, or any other drug for that matter, has its downsides. Using contaminated injection equipment can increase a person's chances of being infected with HIV or other blood-borne diseases. In addition, injection users may be more susceptible to overdose.

Smoking heroin circumvents the potential pitfalls of injecting the drug, while getting it to the brain just as fast. The smoked route takes advantage of the large surface area of the lungs, which have lots of blood vessels to move the drug quickly from the blood to the brain, skipping the liver.

Thinking about the benefits and limitations associated with each of these routes, I would strongly suggest that novices avoid the intravenous route. If a rapid onset of drug effects is sought, then this goal can be readily achieved by snorting or smoking the drug without exposure to the risks associated with injecting. Still, I'm aware that there are plenty of people who continue to inject despite knowledge of the risks and alternatives. For some, the injecting drug-user identity captures the persona they'd like to project. It sets them apart from other drug users, delineates the boundaries of who is down and who is not. Others may have injected for years or decades and are simply more comfortable using this route compared with others. Some who inject have figured out the safest way to do so, under the most sanitary of conditions, and have done so without injury for years.

SET AND SETTING

One of the most frequent questions that I am asked is why people's experiences differ if they've taken the same drug, by the same route, and at the same dose. The short answer is that drug effects are not determined by pharmacology alone. Individual characteristics as well as the environment in which the drug use occurs can greatly influence drug effects. This notion is popularly referred to as "set and setting."

• • •

SET. *SET* REFERS to the individual characteristics of the person who has taken the drug. This factor encompasses everything from people's mood and physiology to their preconceived notions about the substance to the effects they expect to experience from taking it. Drug effects can differ widely, for example, between people who are physically fit, well-rested, and well-nourished and those who are not. Cocaine is renowned for its pleasure-enhancing effects. But the drug can also temporarily disrupt sleep and appetite and, in turn, negatively affect mood. You can greatly increase the odds of experiencing primarily the enjoyable effects of cocaine (and other drugs) if you exercise, eat nutritiously, and get sufficient sleep. Properly looking after yourself contributes to more enjoyable drug effects.

Tolerance is another important aspect related to the concept of set. This is the reduced effectiveness of a drug after repeated administration. For example, a 50-mg dose of heroin taken by a regular user of the drug will not produce the same level of high experienced the first time the person took that dose. The high will be considerably reduced, meaning that the individual might have to take a larger dose to achieve the desired effect. Please don't confuse tolerance with the popular myth that one can never recapture that first high. That's simply not true. Tolerance, however, can be protective. A heroin user who has developed tolerance is less likely to die from an overdose than a nontolerant user is.

SETTING. THE ENVIRONMENT, or the *setting*, in which the drug use occurs can also influence a person's experience. Drug use is expected at festivals, so thoughtful organizers ensure that drug education and medical aid are available to attendees. These services help to make the festival setting more inviting and nurturing to people who use drugs. It's not difficult to see how this environment can lead to a more enjoyable drug experience than can a setting without such support.

DRUG USERS ON THE MARGINS

My time at Parklife got me thinking about drug users who do not have access to such drug-related services and education. The amount of money required to attend a festival, especially when entrance ticket, travel, and food costs are included, is prohibitive for most. Consequently, it is not surprising that the highest drug-related mortality rates in the United States are found in regions, including Appalachia and Oklahoma, with lower rates of university completion and greater economic distress.[12] Attention-grabbing headlines claiming that opioids (or any other drug) are killing people are wrong. Ignorance and poverty are killing people, just as they have for centuries.

In the weeks before Parklife, I spent time in Belfast with my dear friends Buff and Chris. I visited them to learn more about their work. On most nights, the two, along with their team of social workers, can be found walking the city streets, providing destitute drug users everything from a sympathetic ear to drug education to clean needles and syringes. I quickly saw how useful their approach could be for people living in U.S. regions with housing instability and high rates of unemployment.

Buff and Chris are an unlikely pair. They were raised on opposite sides of the sectarian divide that cleaves Northern Ireland to this day. Buff is single and has no children. His solid six-foot-three-inch frame, flattop crew cut, and quiet, wary demeanor can be intimidating, especially to strangers. But he is one of the gentlest individuals I've ever met. Chris is married with children; he's also shorter and thinner than Buff is. His blond hair, blue eyes, easy smile, and boyish face give him an outwardly genial appearance. What bonds them is their unwavering belief that adult drug use should be a protected right no matter your station in life. I saw this firsthand on that cold February night when I hit the streets with their team as they did their rounds. Without hesitation, Katy, a social worker with the team, who stood less than five feet tall, walked into a dark alleyway and spoke gently to a tall, slender man wearing an ill-fitting shirt and worn, unwashed pants. After a few minutes, she returned and explained that because many of their clients are without stable housing, they tend to inject in any secluded place they can find, however unsanitary, including alleyways and public toilets. This behavior, of course, increases the risk of overdoses, as well as of infections and abscesses. In this particular case, she wanted to make sure that client at least had a clean needle and syringe.

In addition to providing users with clean works and naloxone, the Belfast team has been lobbying the city to make available supervised drug-consumption sites. These sites are usually designed to be accessible to drug users who may not be well connected to other health-care services. Many view these facilities as part of a continuum of care for people with addictions, mental illnesses, HIV/AIDS, and hepatitis. At consumption sites, clients are permitted to consume their substance of choice under medical supervision, although the supervision is not overbearing or intrusive. Nonetheless, this component can be crucial if a client experiences a drug overdose, which sometimes occurs in these facilities. Patrons of consumption sites are also provided with clean drug kits, such as needles, syringes, and pipes. Supervised consumption facilities may not be available in Belfast yet, but they are available in a growing number of countries, including Canada and Switzerland.[13]

BACK AT PARKLIFE, I reflected on the stark difference between the setting in which the Belfast users took their drugs and the one in which the festivalgoers did theirs.

I was more convinced than ever that drug-safety testing services ought to be more widely available. Imagine if we could provide the same services for communities around the world and give every user the opportunity to test their drugs to ensure they are safe. Deaths from contaminated drugs would be dramatically reduced.

THE DRAWBACKS OF COMPROMISED DRUG-SAFETY TESTING

I have two important caveats to my praise for the drug-safety testing that is permitted in the United Kingdom. First, the only samples tested are from drugs surrendered by attendees to an "amnesty bin" prior to their entering a festival and from drugs confiscated by the police or medics. In other words, festivalgoers are not allowed to submit samples directly to The Loop, and most attendees do not benefit from the availability of this limited type of testing. Nonetheless, results from some of the surrendered or confiscated drugs are disclosed via social media and flyers posted at the festival. Thus, it's theoretically possible for attendees to learn information about the pills they possess, for example, only if their pills are identical (e.g., same logo, color, size) to pills that have been surrendered and had their test results posted. My second concern about the U.K.'s drug-safety testing is that it is done in collaboration with the police. Involving law enforcement undoubtedly decreases the willingness of many to seek services provided by The Loop or any other such outfit. Think about it. If you were engaged in an illegal activity (e.g., taking MDMA), how likely is it that you would seek the services of an outfit staffed with police? Before you answer, consider my initial interaction with the police at Parklife as I entered the festival alongside white colleagues from The Loop.

Immediately, the police dog handler pointed his animal in my direction only. The K-9 obliged by walking toward me. It sniffed my leg before quickly moving on. In the police officer's eyes, I was now a suspect. This meant I had to undergo enhanced screening to determine whether I had drugs. After several minutes of a humiliating search that required me to remove my shoes and spread my legs before a torch-wielding cop, I was released.

I was pissed. It was clear to me as well as to colleagues from The Loop that I had been singled out because of my appearance. I thought to myself, "This fucking idiot is searching me for drugs when I am headed to the testing facility that has any drug I desire." I also knew that the research literature on drug-detection dogs clearly shows that cues from the handler powerfully affect the dog's actions.[14] If the handler is racist, then guess who will most often be subjected to enhanced searches?

The point is that pleasurable drug-use experiences can be enhanced or diminished depending upon several contextual factors, including the dose taken, the user's level of tolerance, and the setting in which use occurs. Whether we like it or not, recreational drugs are part of our society, and it should be our mission to use this knowledge in support of the health and happiness of drug users. Clearly, part of this mission is to try to keep them safe, not push them into the shadows and force them to risk their lives when there are better alternatives.

4

Drug Addiction
Is Not a Brain Disease

Rarely do we find [people] who willingly engage in hard,
solid thinking. There is an almost universal quest for
easy answers and half-baked solutions.

Martin Luther King Jr.

Nah. This grown-ass man didn't just say that? Incredulous, I stared at the back of Dr. Bob Smith as he quickly walked away from me. At the time, Bob was a senior scientist at NIDA, and we were just seconds away from the start of the tri-annual meeting of the National Advisory Council on Drug Abuse. It was 2013.

This eighteen-member advisory council is composed of experts in scientific fields related to the study of psychoactive drugs. A few members from the general public also serve on the council. In theory, the council provides consultation to Dr. Nora Volkow, NIDA's director, on matters of research direction and funding priorities. In practice, we rubber-stamped Nora's wishes, and everyone knew it.

Nora is an accomplished researcher with hundreds of scientific articles published in some of the field's most prestigious journals. She's best known, perhaps, for her fierce defense of the brain-disease model of drug addiction and her impatience with those who disagree with her on drug- and NIDA-related matters. Many scientists who study drugs, including some at NIDA, believe that she routinely overstates the negative impact that recreational drug use has on the brain and that she essentially ignores any beneficial effects drug use may have. But these scientists don't dare share this perspective with her for fear of repercussions that might negatively impact their ability to obtain grant funding, among other professional perks, from her institute. To put this in perspective, NIDA funds nearly 90 percent of the world's research focused on the drugs discussed in this book. Nora is a kingmaker. She is also seen by some as tyrannical.

Council meetings were attended by NIDA employees, such as Bob. They filled the room like ornaments, there to seem interested, not to be heard unless their comments were in line with the position of their leader. So when Bob quietly said to me, "I really liked your paper on methamphetamine and the brain. But don't tell anyone I said that, because it isn't popular around here," I understood where his fear came from. Still, I was stunned. I couldn't understand a professional in his field being so afraid of his views becoming known that he would ask one of his peers not to tell anyone. That wasn't the behavior of an adult or, at least, not the behavior of any adult I respect. It was the behavior of a child and the expected behavior of NIDA employees. It wasn't as if Bob had said or done anything offensive, unethical, or illegal. He was simply expressing an opinion about a scientific paper on the subject of which he had some expertise.

The paper to which Bob referred had been recently published; it was critical of research studies that drew conclusions far beyond the data collected.[1] A disturbingly large portion of these studies concluded that users of methamphetamine were brain damaged, even though the evidence for this conclusion was weak. Drawing conclusions from scant evidence is a cardinal sin in science. In our paper, we pointed out this and other problems in multiple publications authored by Nora and other researchers.

Prior to the paper's publication, Nora and I were quite cordial. Back in 2007, I helped organize and moderate a series of town hall–style discussions about drugs that featured her speaking to underrepresented high school students and their parents. I suspect I was asked to join her advisory council in part because of our relationship. While we didn't hang like homies, we were definitely friendly. But the views expressed in my paper were undoubtedly interpreted as disloyal. It wasn't exactly the way I had planned on starting my three-year tenure on her advisory council, but it did give me a firsthand look at how scientific indoctrination is accomplished.

Simply put, Nora's perspective—and therefore NIDA's view—is that regular use of recreational drugs damages the brain. So how can any responsible scientist argue otherwise, much less propose that drug use can be beneficial? In this chapter, I will show you. By taking a critical look beyond the pretty pictures produced by brain imaging, I will challenge the notion that regular, responsible drug use causes brain damage. Brain-imaging techniques allow us to view the living human brain while a person is resting or engaged in activities such as complex problem solving. This is a good thing—but even good things have limits. Brain imaging has become so popular that it's difficult to find an article about drugs that doesn't include brain pictures. Many laypeople, as well as some scientists, believe that using brain pictures increases the credibility of the science and findings. It doesn't. You'll see that the sexy images, like the ones Nora and her colleagues so frequently tout to frighten the public, rarely show any actual data. And scientific claims disseminated by authoritative-sounding figures without data are not science.

The major challenge my explanation poses for those with no background in neuroscience is the use of "neuro" technical terms. I have done my best to keep this jargon to a minimum, but a few terms are absolutely necessary. If you hang in there, though, I assure you that no one will ever fool you again with seductive images claiming to show some damaging effect of drugs.

In the 1980s, when I was a high school student, I believed that recreational drug use was unequivocally damaging to the brain. I didn't need any brain images to show me this because I knew drugs produced their effects by acting on the brain. Drugs were bad, I was told. Therefore, drugs were bad for the brain. As might be expected, by this time in my life, I had smoked marijuana on a handful of occasions. I figured that if I didn't inhale deeply and minimized the number of times I smoked, I wouldn't destroy too many brain cells.* I had also drunk alcohol and smoked cigarettes but wasn't smart enough to know that these activities were also recreational drug use.

I saw firsthand what I believed to be the horrors of drug use happening all around me, even in my family. My cousins Michael and Anthony, for example, were sweet, decent boys prior to smoking crack cocaine, or so I thought. Once they developed a predilection for crack, they became homeless criminals, even stealing from their own mother. "Them drugs ain't nothing to play wit," I remember my

---

* Cells in the brain are also called neurons.

mom saying. "Just look at how they ruined ya cousins." We were certain that the drugs had altered their brains and caused them to behave badly. They were the family's embarrassment, and drugs had caused it.

No one stopped to think about the social landscape Michael and Anthony were negotiating before they ever smoked crack. Neither had done particularly well in either middle school or high school. Anthony eventually dropped out and was subjected to chronic unemployment in a job market that routinely discriminated against black people. Michael didn't fare much better, bouncing from one unsuccessful personal relationship and job to another. Neither had the benefit of leaving the neighborhood for any extended period of time—except for when incarcerated—as I did when I joined the air force. The military provided me with a supportive new environment, opportunities to learn new skills, and the experience of success in multiple areas. It would take me decades to understand the role that social and environmental factors played in shaping the lives of my cousins and others we simply labeled as "crackheads," or by some other disparaging term, and discarded as hopeless drug fiends.

So after completing a four-year stint in the military and my undergraduate degree, I began studying neuroscience because I thought only its approach could fix the "drug problem." It was clear to me that the poverty and crime in the resource-poor community from which I came was a direct result of recreational drug use and addiction. I reasoned that if I could stop people from taking drugs, especially by fixing their broken brains, I could fix the poverty and crime in my community.

Do you know the adage "A little knowledge is a dangerous thing?" That was me. I had a little knowledge combined with a lot of ignorance and self-confidence—a recipe for danger. In graduate school, I had learned some neuroanatomy, a little neurochemistry, and a few narrow aspects of how drugs affect the brains of rats. I now had what I believed to be a scientific basis for my high school view that "drugs are bad for your brain," and I thought I was hot stuff. After all, I could now articulate the specific neurons and parts of the brain on which drugs acted.

For example, the nucleus accumbens is a structure located near the front and at the base of our brains. It is rich in the neurotransmitter* dopamine and has been linked to the experience of pleasure. The overly simplistic idea is that when a person experiences pleasure, including from recreational drugs, they do so because the dopamine neurons in the nucleus accumbens have become active. Methamphetamine, for example, causes a release of dopamine in this brain region, and this increased release is correlated with feelings of pleasure. This basic but highly incomplete knowledge had essentially led me, and many, many others, to conclude that there must exist discernible and meaningful brain differences, especially in these dopamine-rich areas, between those who use drugs and those who do not. These differences, I thought, must be the *cause* of addiction and other problems related to drug use.

In 1997, this position was argued eloquently by Dr. Alan Leshner, then director of NIDA, when he published, in the journal *Science*, an influential editorial titled "Addiction Is a Brain Disease, and It Matters."[2] He explained "that addiction is tied to changes in brain structure and function is what makes it, fundamentally, a brain disease." Alan's paper solidified my loyalty to the "drugs are bad for your brain" camp. I was a true believer, and his editorial was my holy writ.

---

* *A chemical signal that allows the communication between neurons. Dopamine is one of many neurotransmitters.*

Around the same time, I began working as a postdoctoral fellow at Yale University under the mentorship of Dr. Elinore McCance-Katz. Today, Ellie is a high-level government official advising Alex Azar, the secretary of health and human services, on matters related to improving behavioral health care in the country. We had very different political leanings, but I saw Ellie as a rigorous scientist who deeply cared about her team. For some reason, our lab's television seemed to always be tuned to the *Jerry Springer Show*, but once we'd had our daily fix of adulterous relationships, chairs flung across the room, fistfights, and stripper poles, we conducted some really good science.

Our research compared the behavioral and physiological effects of intravenous cocaine with those of intravenous cocaethylene, a drug formed in the body when cocaine and alcohol are ingested together. We gave research participants two different doses of cocaine on two separate days, and we did the same with cocaethylene. Placebo (just saline) was given on another day. On each of these five days, we measured the effects of the administered drug on heart rate, blood pressure, and mood to see if there were any differences between the drugs. Back then, many of us thought that cocaethylene was more dangerous than cocaine. Anecdotally, cocaethylene was reported to drastically increase the user's risk of a heart attack or stroke because of its ability to substantially elevate blood pressure and heartbeat.

We completed the study and published our findings in 2000.[3] As it turned out, I was wrong: cocaethylene produces fewer effects on cardiovascular measures than cocaine does, which means it probably carries *less* risk for heart attack or stroke. It wouldn't be the last time I'd be incorrect about a drug-related issue. But it marked a distinct moment in my scientific career because it forced me to reckon with the notion that some of my deeply held beliefs about drugs were simply wrong. Many of these beliefs were not informed by evidence; they were informed by anecdote and conjecture issued by authority figures.

In the above study, we gave several doses of cocaine and coca-ethylene to cocaine addicts, and none behaved inappropriately. They showed up on time for the multiple appointments and complied with our rigid rules for participation. Though their behavior was appropriate throughout, it conflicted with my misguided views about what an addict is and how cocaine alters the behavior of its user. I wrongly thought the study participants would frequently be tardy, miss appointments, behave inappropriately, disregard study procedures, and beg for more cocaine. Boy, was I ignorant. It would take nearly a decade for me to get beyond the baseless and harmful negative stereotypes attributed to drug users.

To conduct this kind of research, NIDA provides researchers with millions of taxpayer dollars each year. It's fair to ask, If cocaine is so neurotoxic,* why would researchers be allowed to give this drug to people? Was the thinking that our participants were addicts, and their brains were already damaged beyond repair, so any further damage caused by the cocaine we were to give would be comparatively negligible? Was it that drug addicts are so damaged that society doesn't need to care about whether they are hurt further? I can't say for sure how either the folks at NIDA or other scientists reconciled this apparent ethical dilemma.

I suspect, however, that many recognized that the dangers associated with these drugs have been greatly exaggerated. I certainly do, but it has taken me nearly three decades of carefully studying drugs to come to this position. It's also important to know that it is difficult to

* *Causing damage or death to neurons*

disentangle politics from science when dealing with a federal organization such as NIDA. Until recently, NIDA's mission statement declared that its goal "is to lead the Nation in bringing the power of science to bear on *drug abuse and addiction*."[4] Drug abuse and addiction are a minority of the many effects produced by drugs, yet this mission statement seems to afford NIDA the tunnel vision to ignore any beneficial drug effects, even though they represent most of our findings.

Undoubtedly, some scientists overemphasize negative effects in order to enhance the public health importance of their articles and grant applications. The greater the perceived problem, the more impactful the research. Other scientists might characterize their behavior as erring on the side of caution. In other words, it is better to highlight *any* potential dangers—even those that are remote—while downplaying potential benefits, including obvious ones. The problem with this type of thinking is that it wrongly assumes that the current lopsided and negative presentation of drug effects on the brain is without serious pitfalls. It's not. Journalists write articles consistent with these half-truths. If you do a quick search of newspaper articles written about any recreational drug, you'll find that almost all focus on negative outcomes. Films and public service announcements employ these distortions in their depictions of drug users. For example, in a popular U.S. antidrug campaign, it is implied that one hit of methamphetamine is enough to cause irrevocable brain damage. For the record, methamphetamine, like its near chemically identical cousin Adderall, is an FDA-approved medication to treat attention-deficit hyperactivity disorder (ADHD). Methamphetamine and Adderall are also approved to treat obesity and narcolepsy, respectively.

Misguided drug policy is often based on these exaggerations. In the 1980s, crack-cocaine use was blamed for everything from extreme violence and high unemployment rates to premature death and child abandonment. Even more frightening, addiction to the drug was said to occur after only one hit, a claim so far from the truth that it's ridiculous. There is no drug that produces addiction after only one use. Drug experts with neuroscience leanings weighed in. "The best way to reduce demand," Yale University psychiatry professor Dr. Frank Gawin* was quoted in *Newsweek* (June 16, 1986) as saying, "would be to have God redesign the human brain to change the way cocaine reacts with certain neurons."

"Neuro" remarks, with no foundation in evidence, made about drugs were pernicious: they helped to shape an environment in which there was an unwarranted and unrealistic goal of eliminating drug use by marginalized citizens, even if it meant trampling on their civil liberties—even if it meant passing absurd laws. In 1986, the U.S. Congress passed legislation setting penalties that were literally one hundred times harsher for crack-trafficking than for powder cocaine–trafficking violations. From a pharmacological perspective, crack is no more harmful than powder cocaine is. They are the same drug. The only difference is that one is smoked (crack) and the other snorted or injected intravenously after being dissolved in liquid. This fact didn't matter; because in 1988, these penalties were extended to first-time offenders and to people who simply possessed crack. Also written into this law was that the United States would become drug free by 1995. Never mind that the removal of recreational drugs from society is both impractical and impossible. You should know that there has never been a drug-free society, it is un-

---

* *An ironic development is that Dr. Gawin, now retired, has become a victim of the war on drugs. For the past twenty years he was prescribed large doses of opioids to control pain associated with Lyme disease. Then suddenly, his physician reduced his dose because of new restrictions placed on prescription opioids. Dr. Gawin complained in a recent interview, "I am in pain . . . I'm depleted. I'm not myself."*[5]

likely that there will ever be one, and almost no one wants to live in such an uninteresting place. Can you imagine a society without alcohol, without caffeine, without antidepressants, without pain medications? Neither can I.

In 2010, the crack-powder law was modified to reduce the sentencing disparity between the two forms of cocaine from 100:1 to 18:1. This change is still insufficient, especially when you consider that more than 80 percent of those sentenced for crack-cocaine offenses are black, despite the fact that most users of the drug are white.

The harmful impact of overinterpreting neuroscience data has not been confined to the United States. The recent actions of Philippine president Rodrigo Duterte represent one extreme example. Just over a year into his presidency, thousands of people accused of using or selling illegal drugs have been killed. Duterte justifies his actions by stating that methamphetamine shrinks the brains of users, and, as a result, these individuals are no longer capable of rehabilitation.[6]

Where would Duterte and others get such foolish ideas? They could simply be reading the scientific literature, especially research funded by NIDA and studies conducted by Nora and her team. Consider a 2016 publication by Nora and her colleagues that warned, "If early voluntary drug use goes undetected and unchecked, the resulting changes in the brain can ultimately erode a person's ability to control the impulse to take addictive drugs."[7] The first clause of this sentence seems to encourage caretakers to be paranoid about any potential drug use, even the nonproblematic recreational use that characterizes the experience of the overwhelming majority who use these drugs. The paranoia this statement will provoke in some parents will likely be far worse than that provoked by any drug effect. The second clause is even more disturbing because it argues that there are inevitable brain changes in response to drug use that cripple the user's self-control. There is absolutely no scientific evidence to justify this statement. Unfortunately, it has far-reaching implications because the paper was published in *The New England Journal of Medicine*, arguably the most widely read medical publication.

What is even more sinister is that similar unsubstantiated claims about alleged long-term brain changes have made their way into the *DSM-5*—the gold-standard for classifying mental disorders: "An important characteristic of substance use disorders is an underlying change in brain circuits that may persist beyond detoxification, particularly in individuals with severe disorders."[8] The wicked insidiousness of this type of sanctioned brainwashing causes countless individuals undue anguish, fearing that their brains are damaged even though they haven't been presented with neuroanatomical proof. Because, there is none.[9]

In order to help minimize the damage caused by baseless neuro-claims regarding drug effects, you have to know how to read the methods and results sections of research papers that are purported to support these claims. I will now show you how to read these sections properly. For starters, you can ignore the introduction and discussion sections of most scientific papers. They usually serve as propaganda instruments to promote the research and ideas of the authors. For those uninterested in the details of this venture, I'll give you the take-home message now: there are virtually no data on humans indicating that responsible recreational drug use causes brain abnormalities in otherwise healthy individuals. Trust me, if this were not the case, I would not so proudly proclaim within these pages my own lifelong recreational drug use.

For those of you who stuck around, I must now provide you with a basic level of understanding about a few commonly used brain-imaging techniques. They can be divided into two categories: structural and functional. Magnetic resonance imaging (MRI) is an example of *structural* imaging. MRI provides high-resolution images of the brain's anatomy, pictures with a degree of focus ideal for detecting structural abnormalities, such as brain tumors or gross neuronal death. MRI procedures are said to be noninvasive because no radioactive chemicals are injected into the person being examined. An important drawback associated with MRI is that it provides no information about how the brain is functioning. It can tell you the *size* of a brain structure but not whether or how well that structure accomplishes a particular task. Simply knowing that my nucleus accumbens is larger than yours doesn't mean that I experience more pleasure than you.

Positron-emission tomography (PET) and functional MRI (fMRI) are examples of *functional* imaging techniques because they can provide brain-activity information that is not available by simply looking at the brain's anatomy. For example, the activity of a specific neurotransmitter can be obtained using a PET scan. Neither PET nor fMRI scans, however, provide information about the anatomy of the brain. But perhaps the most important limitation of PET is that it requires the injection of radioactive chemicals into the person being scanned, although the amount of exposure to harmful radiation is minimal.

Typically, these studies are conducted by recruiting two groups of participants: users of a particular drug and non–drug users. The non–drug users serve as the control group. During a study, each participant's brain is scanned once, and they all complete behavior measures, such as cognitive tests. These scans and measures allow the researchers to determine whether there are behavioral differences between the groups. If so, the brain measures can help to determine the neural (or brain) source of the differences. But because brain images are typically collected at only a single time point for both groups of participants, it's nearly impossible to determine whether drug use caused any observed differences. Any brain differences could have existed before the initiation of drug use. So, as you read the brain-imaging literature, be on guard for the inappropriate use of terms—such as *alterations, atrophy, deterioration,* and *reductions,* among others—that imply that a change has occurred. In order to measure a real change, multiple brain scans need to be completed at different time points. Can you tell if a person's hair style has changed if you have seen the person only once in your life?

STUDY FINDS BRAIN CHANGES IN YOUNG MARIJUANA USERS. That was the title of an article printed in *The Boston Globe* on April 15, 2014. The piece was accompanied by a quote from Dr. Stuart Gitlow, then president of the American Society of Addiction Medicine, who remarked, "It's fairly reasonable to draw the conclusion now that marijuana does alter the structure of the brain . . . and that structural alteration is responsible, at least to some degree, for the cognitive changes we have seen in other studies."

The article and this conclusion were based on a recent MRI study conducted by researchers at Massachusetts General Hospital and Northwestern University.[10] The researchers compared brain sizes of twenty cannabis users with twenty control participants by scanning the brain of each participant *once.* The average age of all research participants was about twenty-one. The cannabis users reported smoking the drug three to four days per week; they also smoked tobacco cigarettes and drank more alcohol than did the controls. The major finding was that, on average, cannabis users had a slightly larger nuclei accumbens and that the amount of reported cannabis use was correlated with accumbens size. The accumbens size differences were small, so small that if the brain scans of all participants were shuffled together into a single stack, it would be nearly impossible to correctly identify the group to which individual scans belonged. But this fact did not stop the investigators from concluding that their results demonstrated "morphometric abnormalities" and suggesting that marijuana exposure "is associated with exposure-dependent alterations" of brain reward structures.

The researchers' interpretations, as well as those in the *Boston Globe*, are inappropriate because brain images were collected at only one time point for both groups of participants. This makes it impossible to determine whether there were any "alterations"; multiple brain scans over time for each participant would be required in order to measure a change. Also important, preexisting brain differences between the two groups cannot be ruled out. In other words, it's possible that the small brain differences were there before the initiation of any drug use. This is a common mistake in the drug literature.

Another frequent oversight is to ignore the influence of tobacco smoking and alcohol use on the findings. In order to disentangle cannabis-related effects from those of tobacco and alcohol, the researchers should have included a third group of participants. This is almost never done in brain-imaging studies. Ideally, this group would have been composed of individuals who reported tobacco and alcohol use but not cannabis use. If the results of this third group were similar to those of the cannabis group, it would suggest that cannabis use was not responsible for the observed findings.

More important, though, there is no way to determine the everyday importance of the small structural differences observed in this study. As you might imagine, there is variability in the size of the nucleus accumbens among individuals. Some people have smaller ones; others have larger ones. Variability in the size range is considered normal, just as there is a range of normal height. Some people are shorter than others, but we would not characterize a five-foot-one woman as evincing "abnormality of stature."

Another critical point is that the study did not include behavioral or cognitive measures. Simply knowing that there is a brain-structure size difference between two groups tells you nothing about the functional integrity of the brain or individual brain structures. For example, it's highly likely that both groups would have performed equally well on a test measuring complex learning and memory or any other domain. Both groups of participants showed up, complied with study procedures, and completed the study. This demonstration of responsibility suggests that even the cannabis users met some basic level of functioning. Still, had the researchers included cognitive tests, for example, specific mental and intellectual information, as well as knowledge about how well brain structures were functioning, could have been deduced. Without carefully measuring a behavior of interest, such as cognition, researchers (and journalists) are often enticed into making unwarranted speculations about the neural basis of behavior. If you don't measure behavior, then you can't comment on behavior.

Unfortunately, most press coverage related to this study was equally irresponsible. Headlines in *The Washington Post* and *Time* declared, Even Casually Smoking Marijuana Can Change Your Brain, Study Finds and Recreational Pot Use Harmful to Young People's Brains, respectively. Such story lines are typical when brain-imaging techniques are used to study drug users. Many of the studies are riddled with vital limitations, and the actual results frequently do not entirely correspond with conclusions drawn by the investigators. What follows are misleading headlines in the general press that are designed to frighten parents, who, in turn, implore their representatives to do something about the "drug problem."

The unjustified alarm about study findings can get even worse when the topic concerns the effects of prenatal drug exposure on subsequent brain functioning of offspring. The popular belief is that prenatal drug exposure inevitably damages the brains of developing fetuses. This opinion is so entrenched that researchers who report findings consistent with this view enjoy less scrutiny of their research prior to publication, especially if brain-imaging data are included. In other words, it is easier to get your findings published if they are in line with the "drugs are bad for the developing fetus" perspective and if brain imaging was used in the study.

I noticed this trend a few years ago while teaching a graduate seminar focused on understanding the impact of prenatal recreational drug exposure on the cognitive functioning of children and adolescents. During the seminar, we critically read and discussed two recent original research articles each session for fifteen weeks. At the end of the semester, rather than requiring students to complete a fifteen-to-twenty-page term paper, I required them to submit a publishable letter to the editor of a scientific journal. The letters had to be in response to recently published articles in the area of prenatal recreational drug exposure, and they were required to incorporate topics, concepts, and principles covered in class. For example, students addressed questions such as, Can causation be determined based on the methods used in the study? Are the conclusions consistent with the data, and are they appropriate? Are the experimental methods appropriate for the stated goals of the study?

To my delight, several students have had their letters to the editor published in some of the finest scientific journals. I was troubled, however, by the empirical confirmation of the bias for publishing research articles that claim to show detrimental effects caused by prenatal drug exposure. Here's just one example. Researchers at University of California, Davis; University of Maryland; and NIDA used fMRI to compare the brain activity of twenty-seven adolescents who were prenatally exposed to multiple drugs with that of twenty controls who were not.[11] Brain activity was measured while the adolescents completed a working-memory test. Prenatal drug exposure included alcohol, tobacco, cocaine, and heroin. Note that most of the drug exposure occurred in the first trimester and then substantially decreased as the pregnancy progressed. This is consistent with most evidence collected on women who have used drugs during pregnancy.

Control participants and prenatal drug-exposed participants exhibited slightly different patterns of brain activation in a few regions, a finding that most likely represents the normal range of human variability in brain activation. All participants performed equally well on the working-memory test. This observation supports the view that brain activation for both groups was normal. Yet the researchers discussed the working-memory findings in surprisingly pathological terms: "The behavioral findings may reflect subtle indications of *altered* [emphasis is mine] attentional and response preparatory skills in the PDE group." I don't understand how equivalent working-memory performance was interpreted as a negative effect for one group (prenatal drug-exposed) but not for the other (control). Unless, of course, the interpretation is driven by bias.

Ultimately, the researchers concluded that their data show "altered neural functioning related to response planning that may reflect less efficient network functioning in youth with PDE." As my student Delon McAllister pointed out in his published letter to the editor,[12] this conclusion extends too far beyond the methods employed and the collected data. That's a nice way of saying that the researchers ignored their own data and told a story that was consistent with their bias. For example, if there were no differences on the behavioral task of interest (the working-memory test), then this precludes statements highlighting such differences. It's like concluding that men are better thinkers than women based on the observation that they both think. Furthermore, neural activation differences alone are insufficient to conclude that one group may be dysfunctional compared with another, especially when the behavior of the two groups does not differ. The differential pattern of effects seen on brain activity in response to the working-memory test is most certainly within the normal range of human variability.

Unfortunately, I don't have enough students to ensure that such bias is minimized in the brain-imaging literature, but I hope the above examples will help you read it with a more critical eye. And I hope this information will decrease the odds of your having the wool pulled over your eyes by less than careful researchers or by those infected with the "drugs are bad" virus.

Another crucial detail that you should know about MRI and fMRI findings is that they are almost never replicated. Replication of findings is a crucial and defining feature of good science. This requirement helps guard against spurious results, unrelated to drug use, from an individual study. Many of these sensationalistic headlines attesting to some new brain finding should be taken with a grain of salt, at least until other researchers have replicated the results.

So far, I have presented examples of how brain-imaging data are misinterpreted and misused. Now I'd like to discuss a study on the other end of the spectrum—one that was extremely well conducted and included appropriate conclusions.[13] The study was conducted by Dr. Chris-Ellyn Johanson, now retired from Wayne State University, her late husband, Dr. Bob Schuster, and other colleagues. I should point out that this study was funded by NIDA and that Bob served as director of NIDA from 1986 to 1992. The point is that it would be a mistake to conclude that all NIDA-funded studies are biased and that everyone affiliated with NIDA is invariably a bad scientist.

Chris-Ellyn and her colleagues used PET-imaging procedures to compare the brains of methamphetamine addicts with nonaddicts. On average, methamphetamine users reported using the drug for ten years; they also reported regular use of other drugs, including alcohol, cocaine, and marijuana. The control participants reported never having used methamphetamine, no history of drug addiction, and no use of illegal drugs except marijuana. The researchers also asked study participants to complete several cognitive tests and compared the performance of the two groups. To help you interpret these findings as well as others, I'd like to provide a few more specific details regarding the PET-imaging procedures used in this study.

A RADIOACTIVELY LABELED CHEMICAL was injected into the bloodstream, and then a computerized scanning device mapped out the relative amounts of the chemical in various brain regions. The radioactively labeled drug binds to specific elements on dopamine neurons. It was therefore possible to see the extent to which binding occurred in all participants.

The researchers published their findings in the journal *Psychopharmacology*. They found that methamphetamine users and control participants performed equally well on most cognitive tests. On measurements of sustained attention and immediate and long-term memory, however, methamphetamine users performed more poorly than controls did. Importantly, though, the methamphetamine users' performance remained within the normal range for their age and educational group. In other words, despite being outperformed by the control group on some tests, the methamphetamine users were cognitively intact. Their cognitive abilities were in the normal range.

Regarding the brain data, on average, dopamine binding in the midbrain was 10 to 15 percent lower in methamphetamine users. It is important to point out, however, that there was considerable overlap in dopamine binding between the two groups of participants. That is, binding for some methamphetamine users was equal to or higher than those of some individuals in the control group. In practical terms, the results mean that if these brain images were shuffled into a single set, experts would not be able to distinguish between the brains of controls and methamphetamine users. The researchers concluded that the functional significance (or everyday importance) of these differences are likely minimal because methamphetamine users' performance on most tests was equal to controls' performance and because there was no relationship between the imaging data and cognitive performance.

As you might imagine, results from this study did not receive major press coverage. Nor were they published in *The New England Journal of Medicine, Nature,* or any other high-profile journal, despite the fact that the study remains one the most rigorously conducted in this area. One reason that it—as well as other well-conducted research with appropriate conclusions—failed to generate media attention is because the researchers chose not to engage in inappropriate alarming speculation about the negative impact of drug use. Instead, the findings were discussed in nonbiased, dispassionate, and cautious terms, the bare-minimum practice we expect from scientists communicating in a scientific journal.

Politicians have long recognized that political and economic currency can be reliably garnered by arousing public fear. The perennial "drug problem" is outstanding in this regard. Today, the problem is opioids; tomorrow, it'll be something else. Votes, money, and influence will go to the politicians who convince the public that there is a problem. Exaggerating drug problems provides politicians with opportunities to be heroes and saviors, even though their solutions rarely work.

So-called drug problems also provide journalists and filmmakers with opportunities to walk on the wild side without getting too close for comfort. Many of these individuals are from the middle class and are intelligently curious but may have little personal experience with specific types of behaviors perceived as morally questionable or risky. Heroin use, for them, would be an example of one such behavior. Joni Mitchell eloquently described this phenomenon in her song "A Case of You" when she sang, "I'm frightened by the devil. And I'm drawn to those ones that ain't afraid." Writing an article or making a film about heroin addiction allows the commentator to walk in the shoes of the addict and then return them once the piece is complete. It's a cheap thrill.

Most drug use—even heroin use—occurs without causing addiction. Yet you would be hard pressed to find a story or a documentary film about an illicit drug, say, crack cocaine or heroin, without its focus being almost entirely on addiction. Why is this? Well, because addiction is sexier than nonaddiction. Who wants to read a piece or watch a film about a person who uses heroin on some evenings and then goes to work as scheduled and handles her other responsibilities without incident? Most people would find this boring, and journalists and filmmakers know this. Disproportionately producing pieces about addiction is a win-win situation for the producers. The paying public eagerly consumes the material, while the journalists and filmmakers pretend to be edgy and hip.

While I find the behavior of most politicians and journalists unacceptable, I also understand that their mistakes are not necessarily malicious. These individuals frequently operate in crisis-ridden environments, where decisions are often made with incomplete information and where deadlines loom large. Blaming drugs for the current crisis promises an easy fix and absolves lots of people from their responsibility for addressing the actual causes, which are usually complex. I see the job of the scientist, in part, as helping to correct the blunders of politicians and journalists. It's one of the reasons that I became a scientist. I enjoy engaging in effortful, deliberate thinking with the goal of accurately and impartially characterizing drug effects on the brain and on behavior.

Unfortunately, this time-honored practice is increasingly being replaced with fearmongering, especially by some scientists who study the effects of drugs on the brain. This occurs even though we hold scientists to higher standards of objectivity than we do politicians or journalists. Misrepresentations of study findings by scientists such as the ones described in this chapter are egregious not only because they impact our treatment of identified drug users; they also contribute to misleading stereotypes and shape callous political rhetoric and harmful policies. On many occasions, Donald Trump has praised Duterte and other barbaric leaders for a "great job" on their handling of drug users and dealers, knowing that their tactics include extrajudicial executions. This is precisely the type of rhetoric that led to passage of the U.S. legislation that set penalties one hundred times harsher for crack- than for powder-cocaine violations.

Today, many find the crack versus powder laws repugnant because they exaggerate the harmful effects of crack and are enforced in a racially discriminatory manner, but few critically examine the role played by the scientific community in propping up the assumptions underlying these laws.

For its part, the scientific community has virtually ignored the shameful racial discrimination that occurs in drug-law enforcement. The researchers themselves are overwhelmingly white and middle class and do not have to live with the consequences of their actions. I don't have this luxury. Every time I look into the faces of my children or go back to the place of my youth, I am forced to confront the devastation that results from the racial discrimination that is so rampant in the application of drug laws and is abetted by arguments poorly grounded in scientific evidence.

We can no longer allow neuro-exaggerations to shape our views on drugs, inform our drug policies, and determine our drug-research funding priorities and directions. The stakes are too high. The human cost is incalculable.

5

Amphetamines: Empathy, Energy, and Ecstasy

Take all away from me, but leave me Ecstasy
And I am richer then than all my Fellow Men

Emily Dickinson

F uck you i will kill you . . . im the one will assassinate you inside the airport. I have more intel inside the Philippine airport to support pres. Duterte." This was the Facebook message that greeted me when I logged onto my computer in the departure lounge at the Ninoy Aquino International Airport in Manila. I was reflecting on the events of the past week and anxiously waiting to board a midnight flight out of the country, several days earlier than my scheduled departure.

I had come to the Philippines at the request of local human-rights organizations—NoBox and the Free Legal Assistance Group Anti–Death Penalty Taskforce—to speak before a Filipino drug-policy forum in Manila. The forum's organizers brought together experts from their country and around the globe to discuss alternatives to the extrajudicial killings of suspected drug users and dealers that had become a disturbing staple of the Philippine drug-control strategy. The organizers asked me to provide a scientific update on the effects of methamphetamine in people, one of my areas of expertise. They knew that I had recently been in Thailand speaking on the same topic. In Thailand, as in the Philippines, methamphetamine use was seen as *the* major drug of concern. The perception of the severity of methamphetamine-related problems was so extreme that the Thai government had passed laws that punish methamphetamine violations nearly ten times more severely than they do heroin violations. This resulted in some of the most egregious prison sentences I had ever seen for simply possessing a drug.

Take the case of Supatta Ruenrurng. On June 7, 2010, Ms. Ruenrurng began serving a twenty-five-year prison sentence for bringing into Thailand from Laos one and a half pills of methamphetamine, worth about five dollars. Combined, the pills contained, at most, 35 mg of the drug, which is a low to moderate dose. To put this in perspective, this amount of methamphetamine is less than the maximum daily approved dose (60 mg) given to children in the United States as part of their treatment for ADHD.

Ms. Ruenrurng's case was not even exceptional. During my time in Thailand, I visited the Udon Thani Central Prison and met at least two dozen other men and women who were serving sentences from twenty-five to thirty years for possessing similar amounts of methamphetamine. As a result of Thailand's harsh drug laws, the country now has the highest female-incarceration rate in the world;[1] more than 80 percent of the women behind bars are there because they had violated a drug law.[2] Overall, Thailand now has the sixth largest prison population worldwide, the vast majority locked up because of drug charges.

Perhaps even more disturbing is that in the early 2000s, Thai drug-law enforcement efforts were more extreme, leading to many more deaths of both users and sellers. In 2003 alone, more than two thousand people were killed without ever having been afforded their day in court.[3]

This is the background context for the unprecedented 2016 conference on drugs in Bangkok in which I participated. The Supreme Court of Thailand, the Ministry of Justice, Her Royal Highness Princess Bajrakitiyabha, and others organized the meeting, bringing together global experts specializing in neuropsychopharmacology, substance-abuse treatment, the judicial system, international drug laws, and law enforcement.

After the two-day meeting, the Thai government was moved to draft new drug laws. This endeavor, according to the Ministry of Justice, would be guided by the best-available empirical evidence, with the goal of balancing compassion for citizens with effective drug-control approaches. Real progress was made.

I was hopeful that the Manila meeting would be fruitful, too, though I was acutely aware of the ignorant statements made by Philippine president Rodrigo Duterte about methamphetamine and the people who use it. On August 17, 2016, less than two months into his presidency, Duterte had proclaimed that "a year or more of shabu [methamphetamine] use would shrink the brain of a person, and therefore he is no longer viable for rehabilitation."[4] Speaking defiantly before an outwardly sympathetic group of Filipino government and police officials, he used the occasion to justify his deadly campaign against suspected drug users and sellers. By the time I was to speak at the Filipino drug forum, nearly five thousand people had been killed extrajudicially as a result of Duterte's bloody drug war.[5]

I was acutely aware that during my presentation I would have to be clear and precise because representatives from the government were expected to be in attendance. I didn't want to be inflammatory or insulting or otherwise reduce their ability to think objectively about methamphetamine or any psychoactive drug. But I wasn't too concerned because I assumed the governmental representatives would be scientists or, at the very least, physicians. This, after all, was a *scientific* forum, an environment where empirical evidence trumps personal anecdotes, even if the anecdotes are touted by the president. I figured I'd simply stick to the facts, to what had been demonstrated under experimental conditions, to what we knew to be true in science.

Besides, it wasn't my first visit to Manila. I had given a number of scientific talks in the country, some of which focused on methamphetamine. And even though some of the talks provoked heated and intense discussions, the best evidence usually determined the positions taken by the attendees. I had no reason to think this talk would be any different.

The fact that the organizers asked me to forward my remarks a week in advance of the forum was a bit strange but not unheard of. In any event, I didn't do it. I later learned that the country's then health secretary Paulyn Jean B. Rosell-Ubial was scheduled to provide a response to my remarks immediately afterward, and she had asked to have access to my comments ahead of time so that she could formulate a rebuttal. Dr. Rosell-Ubial was the official science representative of the Duterte administration. In the end, she didn't show up. Other government officials did attend, however.

At the conclusion of my talk, Agnès Callamard, United Nations special rapporteur on extrajudicial executions, tweeted to her ten thousand or so followers one of my take-home messages: "Prof Carl Hart: there is no evidence Shabu leads to violence or causes brain damage." Of course, it is impossible for an eighty-four-character tweet to capture all the nuance contained in my fifty-five-minute presentation, but it did communicate two of my fundamental conclusions.

Responses to the tweet were swift and relentless. Many supporters of Duterte declared Callamard—and now me, by extension—an enemy of the state. She and I both received threatening emails and social-media messages, and one of the country's most widely read newspapers, *The Manila Times*, published an editorial cartoon mocking me.

In an effort to shed more light on my remarks, I agreed to do a live interview on Rappler Talk, a popular Filipino web-based show. I tried to teach viewers how to separate actual drug effects from the effects stated in baseless or sensationalist claims. Once this distinction is established, I explained, we could begin to solve the real problems people faced in the Philippines, such as high rates of poverty and unemployment, which can lead some desperate people to commit crimes. In my mind, the Rappler interview was a success; I thought it would quell some of the anger ginned up by Callamard's tweet. I couldn't have been more wrong.

After the interview, Duterte himself added his two cents, suggesting that the reason for Callamard's tweet was that we were sleeping together. "She should go (on) a honeymoon with that black guy, the American. I will pay for their travel." Regarding me, specifically, he barked, "That son of a bitch who has gone crazy," because I said the data simply do not justify the claim that methamphetamine use damages or shrinks the brain, certainly not when taken in the typical doses people use.

It is well known that Duterte is a self-admitted long-term opioid user. He claims that his own use is to treat a chronic pain disorder. That may very well be the case, but he doesn't seem to see any inconsistency between his drug use and his vilification of others for theirs.

I sat in the airport lounge, ruminating on the various social-media threats aimed at me. Who, I wondered, would want to kill me for simply stating scientific facts, for providing information that helps to keep people safe?

My mind darted back to June 9, 2013, when I made one of my first national television appearances. As I walked out of the studio, another guest on the show, Billy Murphy, a Baltimore-based attorney who would later represent Freddie Gray's family in their civil lawsuit against the city of Baltimore, handed me his business card.

"Take this, 'cause you'll need it," he said. Billy explained that while he agreed with my perspective, he was concerned for my welfare because others undoubtedly would view my ideas about drugs as dangerous and come after me. I politely took his card out of respect but thought to myself that he was being a bit melodramatic. He wasn't.

In a hypervigilant state, I scoped each person in the lounge. I had always taken great pride in projecting coolness, no matter the situation. In high school, during my years as a deejay, my stage name was Cool Carl. I certainly wasn't feeling cool now. I was intensely aware of the fact that I wasn't in the United States, where at least I knew something about the rules of the game. I was in the Philippines, where a year earlier they had elected Duterte, the same man who boasted about having killed someone with his bare hands prior to becoming president, and who now encouraged vigilante violence against people who used and dealt drugs, and jailed political rivals for voicing opposition to his deadly war.

Earlier in the week, over lunch with a few Filipino politicians, I had learned about the extent of Duterte's human rights violations. Because of the online threats, my lunch hosts expressed concern about my safety. They warned that extrajudicial killings in the Philippines were being ordered and carried out for as little as a hundred dollars per hit. And then there was the eerie interaction I had just had with the airport security guard, who made a beeline toward me as my luggage was being scanned. With a strange smile on his face, he told me that he knew who I was and that I shouldn't disagree with his president. Before I could respond, he was gone.

This shit just got real, I thought. The thought lingered as I read the death threat on my computer. Was I being stalked? Perhaps the message came from that bizarre security guard I had just encountered. Or maybe from someone working on the airport apron. I was scared and wanted to get the hell out of there. I quickly moved away from the huge windows and grabbed a seat against the wall so there would be nothing behind me. My eyes scanned the room.

I put on my headphones in an effort to mask my fear. The reverend James Cleveland was singing, "Nobody told me that the road would be easy," from his classic "I Don't Feel No Ways Tired." The song had a remarkably calming effect.

Robin and I routinely play it, along with other gospel songs, when we create the perfect ambiance for our protected time alone. During these daylong staycations, we shut out the world, take the psychoactive substance of our choice, and intimately enjoy each other.

It was Robin who named the sacredness we experience from these respites. Reared as a devout Catholic, she grew up faithfully partaking in Sunday Mass. Prior to getting married in the Church, we would sometimes attend together, and together we completed the Church's required premarital counseling course. The things we do for love.

These moments are sacred and transcendent. The contributions of the right psychoactive substance to our experience is vital.

The gratefulness that we experience can best be described in religious terms. "Heaven Must Be Like This," by the Ohio Players, paints a picture of the scene. As Robin has said on numerous occasions about our sanctuary, "It is the time that I'm most sure a universal goodness [God] exists." To be clear, she is considerably more optimistic about the existence of God than I am, no matter what drug I have taken. Robin and I are in favor of whatever makes one a better person and, therefore, makes the world a better place for all. I am simply pointing out that we have found, at least from our perspective, what works for us, in line with who we are and with who we are striving to become—more compassionate and humane people.

Without a doubt, amphetamines are our favorite substances to take in our moments of united solitude. Mind you, I was once quite ignorant about amphetamines, about as ignorant as Duterte—well, maybe not as ignorant as Duterte. I didn't know, for example, that this class of drugs includes *d*-amphetamine, the active ingredient in the popular ADHD medication Adderall. Adderall is a combination of amphetamine and *d*-amphetamine mixed salts. The family of amphetamines also includes methamphetamine, MDMA, 2-fluoromethamphetamine (2-FMA), 6-(2-aminopropyl)benzofuran (6-APB), and other compounds.

Because of my ignorance, I avoided amphetamines for much of my life and would not first try them until after I turned forty. It's embarrassing to admit this. Even more embarrassing is what I used to say in response to queries about why I had never taken an amphetamine. My response went something like this: "I don't need 'em, nor do I need any drug." Of course, I didn't *need* to take a drug. Just as I don't *need* to travel by car, train, or airplane. I can simply travel on foot. When traveling over great distances, the other methods are simply far more practical and enjoyable.

I now know that certain drugs, amphetamines definitely included, can enhance pleasure, openness, intimacy, energy, sexual satisfaction, and a range of other experiences normal people routinely seek. Few would balk at using Viagra or Cialis to enhance sexual performance, but many more find it objectionable to use drugs such as amphetamines to improve the sexual experience. I don't entirely know why this is the case, but my guess is that it has something to do with the misguided puritanical values that are so pervasive in our education and that disproportionately regulate our behaviors. I think H. L. Mencken put it best when he defined puritanism as "the haunting fear that someone, somewhere, may be happy."

I remember the first time I took an amphetamine as if it were yesterday. It was my fortieth birthday—October 30, 2006—and I was headed to a NIDA-sponsored meeting.

The long subway ride from DC's airport to Silver Spring was unusually pleasant. The huge smile plastered across my face and the inviting eye contact that greeted fellow passengers undoubtedly freaked out a few. It had been about an hour since I had taken, by mouth, a low dose of methamphetamine. My friend Lorraine, who had a prescription for the drug, had given me a couple of pills as a gift. Lorraine frequently teased me because I was considered an expert on amphetamines but had never actually taken any myself. I sat on the train feeling alert, mentally stimulated, and euphorically serene. "I wish others, too, could experience this feeling," I quietly mused to myself, "and the world might be a happier and better place." At that moment, the world was all right with me.

When the effects had worn off after a few hours, I thought, *That was nice*, then worked out and enjoyed a productive two-day meeting. (Well, it might be a stretch to say that I enjoyed a NIDA meeting.) But I didn't crave the drug or feel an urgent need to take any more. I certainly didn't engage in any unusual behaviors. Hardly the stereotypical picture of a "meth head."

So why is it, then, that Duterte—as well as the general public—has such a radically different view than my own of this drug?

Perhaps it has something to do with so-called public-education campaigns aimed at discouraging methamphetamine use. These campaigns usually show, in graphic and horrifying detail, some poor young person who uses the drug for the first time and then ends up engaging in uncharacteristic acts such as prostitution, or stealing from parents, neglecting basic survival needs, and assaulting strangers for money to buy the drug. At the end of one campaign advertisement, emblazoned on the screen, is "Meth—not even once." We've also seen those infamous "meth mouth" images (showing extreme tooth decay) wrongly presented as a direct consequence of methamphetamine use.

Sure, dryness of the mouth is a side effect of methamphetamine. It's also a side effect of other drugs, including several antidepressants and Adderall. Millions of patients use these medications every day, but there are no reports of dental problems associated with their use. The phenomenon of meth mouth has less to do with the direct pharmacological effects of methamphetamine and more to do with nonpharmacological factors, ranging from poor dental hygiene to media sensationalism. Meth mouth is probably more fiction than fact.

And who hasn't seen the popular U.S. television show *Breaking Bad*? Bryan Cranston, the lead actor, plays a high school chemistry teacher turned methamphetamine manufacturer and dealer. Apparently, playing a meth dealer on TV is enough to make you an expert on the drug and addiction to it, at least that's the perspective viewers of *The Daily Show* were left with after an appearance by Cranston back in 2010.

"Meth is *such* a horrible drug," Cranston spouted to then host Jon Stewart, in part, the actor explained, because it makes users continually pick at their skin in search of imaginary bugs crawling underneath. It took a great deal of inhibitory control for me to continue watching. But I did because Cranston wasn't done. He had a neurochemical explanation for why people become addicted to methamphetamine. "In the beginning," he confidently asserted, "your brain is producing dopamine, along with the drug, which creates the most euphoric high." Users get hooked, he explained, because after a couple of uses, the drug fails to produce euphoria, and then users are only left with their addiction. Stewart, known for his quick, sharp wit and ability to ask tough probing questions, let these claims go unchallenged.

These types of distortions neither prevent nor decrease the use of the drug; nor do they provide any real facts about the effects of methamphetamine. They succeed only in perpetuating false assumptions. What's worse is that this sanctioned public dissemination of ignorance is not limited to methamphetamine. The same "educational" strategy is used to inform the public about other drugs as well.

Swayed by this messaging, the public remains almost entirely ignorant of the fact that all amphetamines are chemical siblings and that methamphetamine produces nearly identical effects to those produced by Adderall.[6] Yeah, you read that correctly.

I know. This statement requires some defense.

I'm not suggesting that people who are currently prescribed Adderall should discontinue its use for fear of inevitable ruinous addiction but, instead, that we should view methamphetamine more as we view d-amphetamine. Methamphetamine, like d-amphetamine, can be used (or, more precisely, is used) to enhance people's well-being and functioning. Remember that methamphetamine and d-amphetamine are both FDA-approved medications to treat ADHD. In addition, methamphetamine is approved to treat obesity and d-amphetamine to treat narcolepsy. In previous decades, both drugs were used successfully as antidepressants. In fact, d-amphetamine is still used—off label—by some psychiatrists in the treatment of depression.

As I said, I, too, once believed that methamphetamine was far more dangerous than *d*-amphetamine use, despite the fact that the chemical structure of these two drugs is nearly identical (see Figure 1). Methamphetamine, compared with *d*-amphetamine, has one added methyl group. In the late 1990s, when I was a PhD student, I was told—and I fully believed—that the addition of the methyl group to methamphetamine made it more lipid soluble (translation: able to enter the brain more rapidly) and therefore more addictive than *d*-amphetamine is. As a budding scientist, I should have known better than to accept claims made about any drug without reviewing the evidence.

FIGURE I

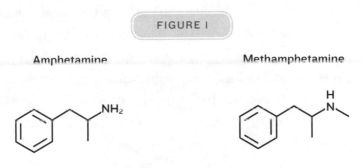

Chemical structure of amphetamine (left)
and methamphetamine (right)

It wasn't until several years after graduate school that my faith-based belief was shattered by evidence—not only from my own research but also by results from studies conducted by other scientists.

In one of my studies, we brought into the lab research participants who regularly used methamphetamine.[7] On separate days, under double-blind conditions, we gave each of them an intranasal dose of methamphetamine, of *d*-amphetamine, or of placebo. We repeated this procedure many times with each person over several days and multiple doses of each drug.

Like *d*-amphetamine, methamphetamine increased our participants' energy and enhanced their ability to focus and concentrate; it also reduced subjective feelings of tiredness and the cognitive disruptions typically brought about by fatigue and sleep deprivation.[8] Both drugs increased blood pressure and the rate at which the heart beats. No doubt these are the effects that justify the continued use of *d*-amphetamine by several nations' militaries, including our own[9]—not to mention the widespread beneficial use of the drug by college students, professionals, and other responsible adults.

And when offered an opportunity to choose either the drugs or varying amounts of money, our research volunteers chose to take *d*-amphetamine on a similar number of occasions as they chose to take methamphetamine. These experienced methamphetamine users could not distinguish between the two drugs. (It is entirely possible that the methyl group enhances methamphetamine's lipid solubility, but this effect appears to be imperceptible to human consumers.)

It is also true that the effects of smoking methamphetamine are more intense than those of swallowing a pill containing *d*-amphetamine. But that increased intensity is due to the route of administration, not the drug itself. Smoking *d*-amphetamine produces nearly identical intense effects as those produced by smoking methamphetamine. The same would be true if both drugs were snorted.

As I left D.C. and traveled home to New York City, I reflected on how I had long participated in misleading the public by hyping the dangers of methamphetamine. For example, in another of my studies, in which I sought to document the powerfully addictive nature of the drug, I found that when given a choice between taking a small hit of meth (10 mg) or one dollar in cash, methamphetamine users chose the drug about half the time.[10] For me, in 2001, this suggested that the drug was especially addictive. But what it really showed was my own ignorance and bias. As I found out in a later study, if I had increased the cash amount to as little as five dollars, the users would have taken the money almost all the time—even though they knew they would have to wait several weeks until the end of the study before getting the cash.[11]

My finding that *d*-amphetamine and methamphetamine produce almost identical effects prompted me to hypothesize that methamphetamine and MDMA would also produce very similar effects. Why not? They had an identical chemical makeup, except for the addition of the methylenedioxy ring attached to MDMA (see Figure 2). And when I studied each of the drugs alone, my research participants seemed to enjoy them equally.

FIGURE 2

Methamphetamine   **3,4-methylenedioxymethamphetamin**

Chemical structure of methamphetamine (left)
and 3,4-methylenedioxymethamphetamine
(a.k.a. MDMA: right)

So I wrote, and was awarded, a grant from NIDA to conduct a double-blind study that would compare the effects of methamphetamine with those of MDMA in the same people. In other words, on separate days, each of the participants would receive an oral dose of methamphetamine, of MDMA, or of placebo. As in the *d*-amphetamine–methamphetamine study, we would repeat this procedure a few times with each person over several days and with two different doses of each drug.

Excited about having been awarded the grant, I glided next door into my colleague Sarah Woolley's office and announced the news and my hypothesis. I was feeling pretty pleased with myself, so I'm certain I came across as reasonably self-assured. Looking back, I realize this wasn't my proudest moment.

Sarah doesn't study drugs. She studies songbirds and the brain mechanisms responsible for the production of the songs they sing. But this didn't stop her from being immediately skeptical. She didn't hold back. Looking deep into my eyes, she asked in her most nurturing tone, "Honey, have you lost your mind?"

Sarah and I joined the psychology department around the same time, a time when the median age of the faculty appeared to be about a hundred and five years old. We shared many laughs about the quirks unique to postretirement-aged academics who refused to retire, no matter how diminished their teaching or social-interaction skills had become. There were only about four or five of us under the age of forty, and none were tenured. It had been nearly twenty years since anyone within the department had earned tenure, and Sarah and I weren't particularly optimistic about our chances. This shared anxiety bonded us; we also bonded over drug conversations.

Much to my chagrin, Sarah now wanted to tell me a thing or two about MDMA and methamphetamine. She went on and on about the personal experiences of her artistic friends, who had used both drugs on numerous occasions; she remarked that no one who had *ever* taken them or who had *observed* someone on them would come away saying that they produce the same effect. Sarah also knew she had the upper hand in this conversation; she knew that at that point in my life I had never taken MDMA.

I got defensive. "Yeah, Sarah," I remarked slightly annoyed, "but you *do* know that anecdote is not evidence." She shot back, "Where is your evidence?" Of course, I didn't have any, at least not yet. I didn't even have an anecdotal experience under my belt. In the end, Sarah would win this argument, just as she has won so many others we've had over the past fifteen-plus years.

I think I got defensive because I felt my expertise was being called into question. She was a friend, yes. Equally important, though, she was a competent scientist, whom I respected. I wanted her to know that I was, too, that I knew my shit. I also wanted her to know that I was right—because I'm competitive and I can be petty. I didn't yet understand how valuable anecdotal drug reports can be, even with their limitations.

I set out to find the truth. In this study, we housed research volunteers for thirteen days.[12] All had used MDMA previously but were informed that they would receive an amphetamine or placebo. In reality, we gave each participant pills containing methamphetamine, MDMA, or placebo on separate days under double-blind conditions. We repeated this with each volunteer over several days and multiple doses of methamphetamine but only one MDMA dose (100 mg).

Both drugs elevated blood pressure and heart rate and decreased appetite. They also produced a substantial amount of euphoria. No surprise here because these are prototypical stimulant effects and undoubtedly a primary reason that people take methamphetamine and MDMA. But only MDMA diminished our participants' ability to focus and concentrate. Methamphetamine, but not MDMA, improved cognitive performance and speech; it also disrupted sleep. In short, methamphetamine and MDMA produce both overlapping and divergent effects.

Still, these results didn't quite explain why the two drugs are viewed so differently. Many people claim that MDMA is like no other amphetamine, especially when it comes to producing feelings of connectedness and emotional openness. They also say that MDMA produces a unique temporary depressive state—colloquially referred to as Suicide Tuesday—in the days following use. We found no evidence of these distinct effects.

Why were my data inconsistent with the stories people tell about these drugs? Were our research methods out of touch with the way people use and experience drugs in the real world? Perhaps. We brought people into a sterile lab, equipped with at least a dozen cameras to record their every move, and asked them to live with three complete strangers for two weeks. Oh yeah, then we gave them MDMA—an illegal drug—and asked them to report their feelings. Not exactly the conditions that would inspire unbridled openness.

I think this is a good place to return to the concept of set and setting. Again, by set, I'm simply referring to the user's mind-set, their preconceived notions about the substance, their expectations of its effects, and their mood and physiology. Setting encompasses the environment: the social, cultural, and physical "place" in which the drug use occurs. These two factors affect all drug experiences and undoubtedly influenced our results. The point is that drug effects are not determined solely by pharmacology (the drug binding to the receptor in the brain). It is biology *and* the environment that determine our drug experiences. This is why knowing only how much dopamine or serotonin was released in response to a drug, for example, when trying to characterize a drug's effect on human behavior or mood, is essentially useless. One must also take into account the environment.

Knowing this, I came up with the idea to try MDMA (100 to 150 mg) and methamphetamine (25 to 50 mg) in the sanctuary of Robin's and my protected time alone. My reasoning was simple. First, this was the ideal intimate setting, one in which I was quite familiar and comfortable. Feelings of anxiety or any other extraneous emotions produced by the environment would be minimized. Second, this environment was always the same, which meant that I could try both drugs on different occasions without worrying about the influence of the environment on the effects of one drug but not the other.

As I had seen in the lab with my research participants, both MDMA and methamphetamine dramatically increased euphoria in Robin and me. What I mean is that we were energized and stimulated and intensely enjoyed each other's company and conversation; we were grateful for our children and the life we had built. Under the influence of MDMA, however, I felt far more empathy, intimacy, and openness than when I had taken methamphetamine.

Another striking difference I have noticed between the two drugs is the phenomenon young people refer to as rolling or the waves. In my experience, rolling is a unique MDMA effect. It can be described as intermittent intense feelings of pleasure, gratitude, and energy. When I'm rolling, I just want to breathe deeply and enjoy it. The simple act of breathing can be extremely pleasurable. I would never have known about this or other unique MDMA-related effects if I had relied exclusively on my lab measures. Before using MDMA, I was too ignorant to even know that I wasn't informed enough to ask the most appropriate research questions. These experiences helped me to more fully appreciate the potential value of anecdotal drug reports.

It's difficult to adequately describe the distinct effects produced by MDMA. I recall a conversation that Lorraine and I once had on the topic back when I was still MDMA naive. I asked her to tell me the difference between MDMA and other amphetamines. She looked at me with a sad, intense, sympathetic look in her eyes and said, "Man, if you don't know, you'll never know." I snapped back, "What kind of fucking response is that?" I now get it.

Still, I'm left with the task of explaining this difference in terms that non–MDMA users might understand. Perhaps this music analogy will help. In 2015, on a cold November night in a Liverpool hotel, my musician friend Steven pulled up a YouTube video of Al Green singing "Jesus Is Waiting." This was a recording of Green's April 6, 1974, *Soul Train* performance. At the time, *Soul Train* was the most popular black American music and dance television program. But on that day, Green would transport the *Soul Train* dancers to the black church.

Standing before the audience, with his arm in a sling around his neck, he began with the Lord's Prayer. With his eyes closed for much of the song and sweat streaming down his face, Green gave the most transcendent performance I had ever seen on *Soul Train* or anywhere else. He reminded me exactly why I so value our shared sacred space: "[I] said if ya, if ya broken down. Jesus is waiting. Don't let yourself down."

Back in the 1970s, Green was one of my mom's favorite artists, so I had heard this song many times. It was part of the soundtrack of my childhood and is still one of my favorite songs. But I can say with absolute certainty that none of the other recordings of "Jesus Is Waiting" comes close to moving me the way that *Soul Train* performance does. That's the difference between MDMA and methamphetamine.

As I reconsider my data from our study that compared methamphetamine with MDMA in the lab, I'm haunted by the many things I did wrong. Had we allowed our participants to take each drug with an intimate partner in a private setting, I think we would have been in a better position to document more clear differences between the effects produced by each of the drugs. Of course, we should have also included direct measures of empathy, intimacy, and openness, among others that more closely correspond to the unique effects that many people report when under the influence of MDMA. At present, the measures typically used in lab studies fall short at capturing some of the most crucial aspects of recreational drug use.

It took me nearly twenty years and dozens of scientific publications in the area of neuropsychopharmacology to recognize my own biases against amphetamines. I can only hope that you don't require as much time and scientific activity in order to understand why reasonable adults might use this class of drugs.

And I hope that this knowledge engenders less judgment against and greater empathy for people who use amphetamines.

As I looked out of the window of the Seoul-bound Korean Air flight as it took off from the Manila airport, I felt a tremendous amount of ambivalence. I was relieved, of course, because I no longer felt my life was in danger, but I was also profoundly sad for the people of the Philippines, especially the poor. They were the primary targets of Duterte's war on methamphetamine. It also wasn't lost on me that in medicine, methamphetamine is used to improve the lives of patients. Recreational users take it to feel good and increase energy. In short, amphetamines help to make people feel better. How can we be against people pursuing happiness? As I settled into the flight, The Isley Brothers were singing into my headphones, "Dress me up for battle, when all I want is peace . . . Nation after nation, turning into beast."

6

Novel Psychoactive Substances: Searching for a Pure Bliss

The thing is this:
You got to have fun while you're fightin' for freedom,
'cause you don't always win.

Molly Ivins

I gotta say," Robin whispered conspiratorially, "this is my favorite drug." We were alone, in bed, listening to Bill Withers sing "Grandma's Hand." We were decompressing, reflecting on our handling of some recent difficult events. Two hours earlier, we had each taken a 150 mg dose of 6-APB, a novel psychoactive substance structurally similar to MDMA (Figure 3). Its peak effects were kicking in and prompted Robin's disclosure. She described how 6-APB helped her to embrace vulnerability and not fear it. How it focused her attention on the things that *really* mattered and not on insignificant daily annoyances. Unhurriedly, with a great deal of thought-

fulness in her voice, she remarked, "The 6-APB experience . . .
it's . . . nurturing . . . and protective . . . um . . . just like 'Grandma's
Hand.' "

FIGURE 3

3,4-methylenedioxymethamphetamin   6-[2-Aminopropyl]benzofuran

Chemical structure of 3,4-methylenedioxymethamphetamine
(a.k.a. MDMA: left) and 6-[2-Aminopropyl]benzofuran
(a.k.a. 6-APB: right)

"Mm-hmm," I concurred, while listening patiently and intently
(not my strongest virtues). But under the influence of 6-APB, I was
pleasantly content to shut my mouth and eyes and open my mind
and ears. The grief-filled, unproductive ruminations usually housed
in my head were gone. The simple act of breathing deeply brought
me unadulterated joy. Robin continued, "6-APB . . . a pure bliss,
times six!" Multiplied by six, she said, because "the high is incredibly
gentle and long-lasting." It was just before midnight on February 22,
2019.

MY INTRODUCTION TO NOVEL
PSYCHOACTIVE SUBSTANCES

When I use the phrase "novel psychoactive substance," I am referring to a catchall categorization that includes everything from synthetic cannabinoids to synthetic stimulants to a host of other little-known mood-altering compounds. A unifying feature of these chemicals is that each resembles a "classic" or "established" drug, such as amphetamine or marijuana, in terms of chemical structure and psychological effects produced. Also, many are relatively new and unknown to authorities. This means that they may be legally available through the internet and other sources.

Before 2016, I wasn't hip to 6-APB or several other popular novel psychoactive substances. This would change as a result of two separate, unexpected encounters with Catalans.

"Carl Hart . . . Carl Hart!" I heard a woman's Spanish-accented voice yell from afar, as I walked out of the United Nations building in New York City and made a beeline toward a waiting taxi. It was April 19, 2016, and I had just given a presentation at the United Nations General Assembly Special Session on Drugs (commonly known as UNGASS). The meeting was held to assess cooperation between nations in carrying out an integrated and balanced approach to dealing with global drug issues and to make recommendations in the service of this goal. In reality, it was an international campaign rally to promote slogans in line with ridding the world of the drugs discussed in this book. What a colossal waste of time.

The woman, whose name I learned is Araceli, ran over and cut me off before I could open the door. With excitement in her big, soft dark eyes, she blurted out, "I came here to meet you!" Araceli had come from Barcelona, where she was working for a nonprofit called Energy Control. When I told her that I was unfamiliar with this organization, she was unfazed and explained that they provide free, anonymous drug-safety testing services to users of illicit drugs. In this way, drug consumption is a less precarious activity because consumers are informed about the chemical constituents contained in their substances. "Brilliant," I thought.

But the fact that I had never been to the Catalan capital was a bit too much for her to handle. Her huge eyes were filled with sincere pity. "How is it possible that a drug expert has never been to Barcelona?" she asked. She let me know that if the Spanish were as arrogant as the Americans, then Spain, and Barcelona in particular, would be known around the globe as the "Center of the Drug Universe" because of the innovative and humane drug policies practiced in her country. I also learned from Araceli that the Spanish had decriminalized all drugs as early as 1973, long before the Portuguese did in 2000. Meeting Araceli would more than make up for the time I wasted at the UNGASS fiasco.

Later that year in August, at the Boom Festival in Portugal, I met another Catalan, José. Like Araceli, José was working at Energy Control; he, too, emphasized that Barcelona was the place to be for anyone seriously interested in drugs. It was José who first introduced me to 6-APB. "It's a better MDMA," I remember him saying. "It caresses your soul so you can do the same for others." Ultimately, José and Araceli joined forces to convince me to visit Barcelona.

In April 2018, two years after UNGASS, I arrived in the City of Counts for the first time. Ostensibly, I was there to complete a grant application with José. We were seeking funds to organize a conference at Columbia that would offer solutions to stem the tide of drug-related deaths in the United States. A fundamental guiding principle of our meeting, in contrast to that of a typical academic forum on drug use, was that consumption of psychoactive substances is a normal human pastime, as old as humankind itself. It would be foolish to expect our species not to use drugs. The task of responsible governments is to balance the natural human desire to alter one's mood with the public's health and safety. One of our conference's goals was to spotlight anonymous drug-safety testing services. These services, in our view, hadn't received sufficient academic attention and their use was virtually absent in the United States. This sorry situation exists despite the fact that drug-safety testing has been empirically demonstrated to reduce the number of users exposed to harmful effects of adulterated substances.

Another goal of our conference was to focus attention on evidence showing that overly restrictive drug laws actually contribute to the proliferation of novel psychoactive substances. Such proliferation can also put users' health at risk. I know many people believe that once a drug is banned, demand for that drug (or its desired effects) will dissipate. Nothing could be further from the truth. There will always be a demand for commodities that enhance joy and mitigate human suffering. A number of novel psychoactive substances, 6-APB included, are outstanding in this regard. This is one reason they continue to proliferate. To circumvent harsh legal restrictions placed on so-called classic drugs, illicit manufacturers simply synthesize and sell novel psychoactive substances as alternatives. For example, mephedrone is occasionally sold as an alternative to MDMA because the two drugs produce similar effects, although effects produced by mephedrone are shorter in duration.[1]

Still, as is the case with any new drug, many associated risks may not be as well known as those associated with the classic substance.

Also, deceptively selling a new drug as an established one can lead to harmful effects, especially if the pharmacological profiles of the two differ in a meaningful way. For example, as we've seen, if an unsuspecting person consumes a large amount of carfentanil, a novel psychoactive substance that is sometimes substituted for heroin, the consequences can be fatal. This is a too frequent, but predictable and preventable, outcome caused by heroin prohibition.

Promptly after completing the grant application, I began my Barcelona drug education. Araceli and José gave me a tour of Energy Control and explained the inner workings of its drug-safety testing. Each year, they analyze thousands of samples using gas chromatography and notify anonymous users of the detailed chemical results. Employees at Energy Control also provide basic drug-safety education, including information about dosing and specific drug combinations. All is available to Spanish citizens free of charge. Compared with most drug-safety testing I had seen in other countries, the Energy Control model was more comprehensive. I was stunned by the elegant simplicity of their approach. The primary intent was to keep users safe, not infantilize them, and to respect their autonomy.

Afterward, we visited with multiple friends and associates. All worked in or around drugs. Some were chemists, pharmacologists, and physicians—others were activists and harm-reduction workers. Each had a distinct expertise and was delighted to speak about it at length. Pablo, for example, was an inventor. One of his many devices automatically delivers an injection of naloxone to the user when blood-oxygen levels drop below a certain point. The lifesaving potential of this apparatus for anyone undergoing an opioid overdose is obvious.

But the thing that I remember most about these visits is that each individual had a personal drug stash. Unlike in the United States, in Spain, having a personal stash wasn't a crime. Drugs were decriminalized. What's more, people's private inventories were stocked with pharmaceuticals that had been tested at Energy Control. "So, this is what freedom looks like" was the dominant thought echoing in my head.

"Can I put a hex on you?" Catalina asked invitingly, as she dangled before me a clear plastic baggie filled with snow-white powder. It wasn't obvious to me whether she was joking about casting a spell or offering me a drug or both. The dumb look on my face revealed my confusion. She laughed and enunciated each word slowly: "Would . . . you . . . like . . . some . . . hex?" Mind you, at that point, I hadn't even heard of hex, let alone thought about being faced with this question. "Huh?" I responded. "What is it?"

Catalina explained that the drug is called hexedrone, hex for short. It's a synthetic derivative of cathinone. Structurally similar to amphetamine, cathinone is the main psychoactive component found in the East African shrub khat. Some chew the leaves of khat to obtain stimulant effects. Over the past fifteen years or so, the popularity of synthetic cathinones has increased, in part, because these chemicals engender mood-enhancing effects similar to those produced by drugs such as amphetamine, cocaine, and MDMA. N-Ethylpentedrone, 3,4-methylenedioxypyrovalerone (MDPV), methylone, and mephedrone are just a few synthetic cathinones used recreationally.

In the United States, these drugs are typically referred to as "bath salts" because they were once disguised and sold as such in order to get around prohibitory drugs laws. You might recall the case of the Miami cannibal in which bath salts were initially blamed. On May 26, 2012, Rudy Eugene, a thirty-one-year-old, emotionally unstable man, suddenly attacked Ronald Poppo, a sixty-five-year-old homeless man, on a busy Miami road.[2] During the near twenty-minute assault, Eugene repeatedly bit Poppo's face. By the time police arrived and shot Eugene dead, Poppo had lost an eye and half his face.

Armando Aguilar, president of the local police union, immediately speculated that "bath salts" triggered the gruesome attack. These drugs "[cause users] to go completely insane and become very violent," he said. To bolster his case, Aguilar told ABC News about multiple other incidents he claimed to know involved bath salts. "It took fifteen officers to stop him, and as he was being tasered, he was begging them to shoot and kill him," Aguilar alleged about an incident following a music festival. Paul Adams, an emergency-department physician, backed up Aguilar's claims about the propensity of these drugs to produce superhuman strength. "You can call it the new LSD," Adams told ABC News. "They [patients] are not rational, very aggressive, and are stronger than they usually are. In the emergency room, it usually takes four to five people to control them, and we have had a couple of people breaking out of restraints."

Putting aside for the moment that neither Aguilar nor Adams possessed toxicology results confirming that Eugene had consumed synthetic cathinones (or any other drug), the evidence from research investigating the effects of these drugs is spectacularly inconsistent with the assertions made by these trusted public servants. Synthetic cathinones produce stimulant-like effects. No recreational drug creates superhuman strength or incites the type of violence described by Aguilar and Adams.

Unfortunately, these sobering facts didn't stop the ensuing media frenzy. Lurid headlines screamed, NEW "BATH SALTS" ZOMBIE-DRUG MAKES AMERICANS EAT EACH OTHER and FACE-EATING ATTACK

POSSIBLY PROMPTED BY 'BATH SALTS,' AUTHORITIES SUSPECT.[3] The content contained in these articles was even worse. Natashia Swalve and Ruth DeFoster conducted a critical review of major press coverage surrounding the incident.[4] They found that this coverage was replete with sensational descriptions of the attack and of purported effects produced by synthetic cathinones. Multiple reports claimed that use of bath salts had reached epidemic proportions—it hadn't—and called for harsher legislation related to these drugs. Evidence from science on synthetic cathinones was conspicuously absent.

Then, on June 27, 2012, one month after the attack, Eugene's toxicology results were made public. No synthetic cathinones were found in his system. The only drug found was Δ^9-tetrahydrocannabinol (THC), the major psychoactive component in marijuana. But the minute amounts contained in Eugene's body suggest he hadn't smoked on the day of the attack. The truth is, we still don't know what caused Eugene to behave so violently. It wasn't drugs, though. How about mental-health issues or overzealous religiosity, as has been suggested by some? Perhaps. But the evidence in support of these perspectives, too, is limited.

We did learn, though, thanks to the investigative reporting of Frank Owen, that Aguilar likely peddled the drug-induced cannibal story as a distraction. According to Owen's theory, rather than risk that the killing of Mr. Eugene, who was black, by officer José Ramirez, who is Latino, become a public-relations problem for the Miami Police Department, Aguilar "decided to bury the racial angle by feeding local reporters an alternative narrative that would prove irresistible: A flesh-eating monster high on a sinister new drug called bath salts devoured a homeless man's face."

It saddens me that the old cliché "the crazed Negro drug fiend" continues to be believed, even to this day. That's one reason law-enforcement officials such as Aguilar tout incredible drug stories so brazenly.

As far as the link between the attack and synthetic cathinones was concerned, the truth was now out; but the damage had already been done. As one example, the federal government passed the Synthetic Drug Abuse Prevention Act, which banned a number of novel psychoactive substances, including mephedrone, on the very same day Eugene's toxicology findings were released. And even though synthetic cathinones were not involved in the Miami attack, subsequent news stories still blamed the drug. CBS News published one such article on April 2, 2015, nearly three years after the facts were known.[5] Perhaps journalist Jacob Sullum, who has written extensively and responsibly on the topic, put it best: "The legend of zombie drugs shambles on with the help of panic-promoting journalists who know a good story when they see one and do not much care whether it's true."[6]

Taking into account my own knowledge regarding synthetic cathinones, supplemented with the information Catalina provided, I enthusiastically accepted her offer: "Let's do it!" She meticulously weighed out the hex in several lines that ranged from a 30 to 50 mg dose. Within fifteen minutes, the pleasurable effects were strikingly apparent. I felt euphoric, energetic, clearheaded, and highly social. It was niiiiiice; it felt like cocaine. Importantly, the hex-related effects weren't overwhelming or disorienting, so they facilitated substantive discussions about hex, as well as other drugs. When the effects had worn off after a couple of hours, I didn't feel any mood disruptions or other residual effects that might signal a cause for concern. The experience was unequivocally wonderful. I certainly didn't have a desire to eat anyone's face.

In fact, it was so enjoyable that I tried other synthetic cathinones, including N-Ethylpentedrone and 2-methylmethcathinone, while in Barcelona. The others also produced pleasing effects, but hex left an enduring impression on me. I now include hex among the drugs I might want to take immediately before attending some awful required social event, such as an academic reception or an annual departmental holiday party. Hex, like alcohol, increases affability, euphoria, and energy—all conducive to a party atmosphere. It's ironic that I have to think about whether hex would be acceptable, even as colleagues and staff openly overindulge with alcohol during such events. Why should we be limited to alcohol in these settings when many other drugs also increase sociability?

I also smoked opium (obviously not a novel psychoactive substance) for the first time in Barcelona and felt remarkably serene, mellow, and contemplative. I felt fortunate to have been welcomed into this community of responsible professionals, who also happen to use drugs. Some were parents, too—and their children, by the way, looked healthy and happy, loved and well cared for. These grown-up drug users spoke candidly about their *current*, not past, use. And our conversations weren't stifled by tension or weirdness, as is often the case when openly discussing the favorable aspects of drug use. Instead, I was treated to a refreshingly free exchange of ideas and information on my favorite topic, not to mention the pleasure of being lifted by the psychoactive treats. It was exhilarating.

BORN IN THE U.S.A.

I also couldn't escape the sobering thought that because I lived in the United States, the enviable psychoactive pursuits of this com-

munity were taking place mostly outside my field of vision and without me.

When I returned home, this feeling was amplified. Synthetic
cannabinoids were now in the press's crosshairs; they were the new
drugs to be feared. An article published in *The New York Times* on
May 20, 2018, noted that "eradicating synthetic cannabinoids has
been a struggle as their use persists."[7] Any thinking person knows
that it is virtually impossible to eliminate all synthetic cannabinoids.
This class of drugs is extremely large and contains multiple medications that are currently used therapeutically, including cannabidiol
(CBD), which is used to treat seizures. Exasperated, I thought to
myself, "Here we go again."

The more I think about the synthetic cannabinoid story, the
more I am convinced that it highlights much that is wrong with the
American approach to drug policy. Synthetic cannabinoids are also
the fastest growing and largest group of novel psychoactive substances, so I'd feel remiss if I didn't discuss this group of drugs in
some detail.

Initially, these chemicals were synthesized by scientists, such
as John W. Huffman and others, for the purpose of studying the
endogenous cannabinoid (endocannabinoid) system. Huffman's research, for example, sought to better understand how this class of
drugs could be used in medicine. The immense importance of this
line of research isn't difficult to see, given that the endocannabinoid
system is a widespread network located throughout the body. By
this I mean that this family of chemicals and their corresponding
receptors are found in many brain structures and in other areas outside the brain. It's involved in many basic functions, including appetite, immunity, memory, mood, pain, and sleep.

Scientists weren't the only ones interested in these compounds. Marijuana users were interested in them, too. Like THC, many of these drugs stimulate endocannabinoid receptors in the brain. Some, when inhaled, produce marijuana-like effects, including euphoria and relaxation. That's why they are sometimes referred to as synthetic marijuana. Recall, too, that recreational marijuana was banned everywhere in the United States until 2014, when adult use was legalized in only two states (Colorado and Washington).

By the early 2000s, word was out that synthetic cannabinoids were legal and available to anyone in the know. They were sold in "head shops," in convenience stores, and through the internet. They were marketed as natural herbal incense or potpourri under various brand names, such as Spice or K2. Most were undetected by conventional urine drug tests, an appealing feature for anyone subjected to random drug screens. All these developments contributed to the appeal of synthetic cannabinoids for people seeking a legal alterative to marijuana.

Then, in 2011, the legal status of synthetic cannabinoids began to change. Authorities in the United States banned five specific compounds (Table 1). This ban began a cat-and-mouse game between authorities and drug producers, resulting in the rapid introduction of slightly modified chemicals to circumvent existing laws. It works like this: law enforcement detects a new synthetic cannabinoid in the illicit market; that cannabinoid is then banned, followed by a proliferation of new replacements, which are usually more potent and potentially more dangerous to the uninformed consumer. With each successive year, the number of banned and new synthetic cannabinoids continues to grow.

TABLE 1

Synthetic Cannabinoid	Year Banned
Cannabicyclohexanol; CP-47,497; JWH-018; JWH-073; and JWH-200	2011
AM-2201, AM-694, JWH-019, JWH-081, JWH-122, JWH-203, JWH-250, JWH-398, SR-18, and SR-19	2012
APINACA, UR-144, and XLR-11	2013
5F-PB-22, AB-FUBINACA, ADB-PINACA, and PB-22	2014
AB-CHMINACA, AB-PINACA, and THJ-2201	2015
ADB-CHMINACA	2016
5F-ADB, 5F-AMB, 5F-APINACA, ADB-FUBINACA, AMB-FUBINACA, MDMB-CHMICA, and MDMB-FUBINACA	2017
4-CN-CUMYL-BUTINACA, 5F-AB-PINACA, 5F-CUMYL-P7AICA, 5F-EDMB-PINACA, 5F-MDMB-PICA, FUB-144, FUB-AKB48, MAB-CHMINACA, MMB-CHMICA, NM2201, and SGT-25	2018

A list of synthetic cannabinoids and
the year they were banned in the United States

The cat-and-mouse game can have real consequences on the health of consumers of synthetic cannabinoid products. Consider the following. Before 2011, JWH-018 was most likely the active ingredient in products sold as K2. When smoked, it produced effects similar to those of THC, although at lower doses. In 2011, JWH-018 was banned, prompting manufacturers to replace it with a less known, more potent synthetic cannabinoid. In some cases, multiple drugs were included as replacements. Hence, after JWH-018 was banned, buyers of K2 most likely received a different drug or multiple drugs. To make matters worse, the package labeling on many of these products did not always accurately reflect the contents, including dose and drug. It's not difficult to see how this could have led to unexpected, malign effects in unsuspecting consumers.

A good case in point is the events of July 12, 2016, in New York City.[8] On this day, thirty-three people in a predominantly black Brooklyn neighborhood were reported to be stupendously intoxicated after consuming what was referred to as synthetic marijuana. Some of these individuals became debilitated and disoriented and even temporarily lost consciousness, but fortunately, no one died. Meanwhile, local and national media headlines included SYNTHETIC MARIJUANA OVERDOSE TURNS DOZENS INTO "ZOMBIES" IN NYC.[9] Accompanying stories and videos dramatized the extraordinary potency and ill effects of K2. Each was peppered with sensational quotes such as this one from an article in The New York Times: "It's like a scene out of a zombie movie, a horrible scene."[10] The conspicuous moralism dehumanizing users was palpable.

And predictably, virtually all these pieces were devoid of any useful information that would enhance the health and safety of synthetic cannabinoid users. For example, not one article confirmed that a synthetic cannabinoid had indeed been ingested. Not one reported the actual contents contained in the products the victims were alleged to have consumed. Not one mentioned that the observed ill effects could have been caused by other substances or some other factor. This point is particularly important because most of the victims had been observed next to a local methadone clinic, where some may have been patients. Obviously, the combined effects of the opioid medication with other drugs could have been a contributing factor in the reported adverse effects.

I was frustrated by such irresponsible reporting and patent neglect for public health. I went on a local news program calling for city health officials to retrieve the alleged products and test them in order to determine their constituents. I also called for officials to obtain biological assays (blood and urine) from patients who were transported to the hospital to see if this information corresponded with results from the tested products. In this way, specific possible causes could have been carefully investigated and findings could have been widely publicized. Members of the press, the local community, and the broader drug-using community, among others, could have been alerted to prevent further harms to those who might otherwise consume the causal agent. This didn't happen, at least not initially.

Several months passed before the public was provided with any useful information regarding a potential culprit in the so-called zombie outbreak. On December 14, 2016, five months after sensationalistic zombie stories appeared in the popular press, an article published in *The New York Times* reported the villain to be a new, more potent synthetic cannabinoid.[11] This perspective was taken from the conclusions drawn by researchers in a recently published article in *The New England Journal of Medicine*.[12] The researchers obtained and tested blood and urine samples from eight of thirty-three individuals who experienced adverse effects. The researchers also tested a sample of the herbal "incense" product "AK-47 24 Karat Gold," which was claimed to have been the item responsible for the ill effects. The findings revealed that the synthetic cannabinoid

AMB-FUBINACA was identified in all eight individually tested packets of AK-47 24 Karat Gold and that its metabolite (by-product of the body breaking down a chemical) was found in the blood of all eight individuals. Importantly, the amount of AMB-FUBINACA contained in individual AK-47 24 Karat Gold packets varied, ranging from 14 to 25 mg/g. In addition, half of the tested patients had other drugs in their system, including an antidepressant, an antihistamine, a benzodiazepine, and opioid medications.

Back in 2016, when the Brooklyn incident occurred, AMB-FUBINACA was not yet banned. For this reason, sellers probably included it in their products as a replacement for a recently banned synthetic cannabinoid. The problem is that AMB-FUBINACA is considerably more potent than THC—and even more potent than JWH-018—meaning that far less of this substance is needed to produce effects, including unfavorable ones. Now that it's banned, less well-known and likely more potent replacements will fill the void.

That's why the government's knee-jerk response to ban any new psychoactive substance invariably leads to more unknown substances in the illicit market. This pattern has been repeatedly shown to jeopardize the health of people simply seeking to alter their consciousness. Note also that most users of synthetic cannabinoids consume these substances seeking a marijuana-like high and that serious adverse effects are rarely associated with adult marijuana use. Furthermore, an outbreak of negative health reactions to synthetic cannabinoids—like that which has been reported in several states, including Connecticut, Illinois, Maryland, and New York—is virtually unheard of in states where marijuana is legal. If you were serious about reducing problems associated with illicit synthetic cannabinoids, you would push for the expansion of legalized recreational marijuana.

It's utterly disheartening to know that regular, decent people's health is unnecessarily placed at risk because of dishonest, callous leaders. When it comes to formulating drug policy, too frequently lawmakers are allowed to simply ignore evidence—and even make up their own. I can't count the number of times I've seen shameless politicians use this strategy to mislead the public about drugs.

Take for example, former speaker of the house John Boehner, who opposed marijuana legalization for his entire thirty-year political career. Back in 2011, he wrote a constituent to say "research shows that marijuana use in its raw form is harmful" and that he was "unalterably opposed to the legalization of marijuana."[13] In 2018, following Boehner's 2015 resignation from Congress, he joined the board of Acreage Holdings, a Canadian firm that is the largest multi-state owner of cannabis licenses and assets in the United States. As you might've guessed, Boehner no longer opposes marijuana legalization. Now he's an advocate. Now he believes that laws prohibiting the substance are "way out of step."

Boehner is a hypocrite. Don't misunderstand me, I think marijuana should be legalized nationwide. My position is unequivocal. What's more, I have a tremendous amount of respect and admiration for those who are able to change their minds in the presence of new and better evidence. This is called cognitive flexibility, a hallmark of intelligence. Boehner's newfound position, however, was motivated by greed. He doesn't seem to give one fuck about the extensive harms caused by the prohibitory policies he once supported. These policies compromise the health of synthetic cannabinoid users and facilitate racism in law enforcement. "I don't have any regrets at all," Boehner told National Public Radio. Astonishingly, he said, "The whole criminal justice part of this, frankly, it never crossed my mind."[14]

Recently, the dishonorable behavior of those in leadership positions hit close to home. On Valentine's Day 2019, my wife and I spent the entire morning at our son Malakai's school. We were there because a video had been posted on social media in which another student called Malakai a nigger. Both the head of school and principal expressed shock at their student's use of the *n*-word. But when we asked about their plans to address the issue, they hadn't yet come up with one. Then when we asked to see the video, they refused.

These people didn't exactly inspire our confidence. On numerous previous occasions, we had brought to their attention far worse race-based transgressions directed at our son. We were politely dismissed each time. For just one example among many, the security staff routinely singled out Malakai and other black boys for photo identification inspections before entering the campus. Such ID checks were highly unusual for this close-knit school community—one in which we had been members for nearly twenty years and one that prides itself on the honor system and high ethical standards—and where the tuition is more than fifty thousand dollars per year.

An incident that occurred back in May 2018 is emblematic of the double standards my son was frequently subjected to by the school itself. He, along with a few other black boys—still wearing their school track uniforms—and their black track coach, had just returned from a competition held at another school and were headed to the locker room to retrieve their ID cards and other school-related items, including their homework. They also wanted to change out of their track uniforms into everyday gear. To their surprise, however, they were stopped at the security booth and denied access to the campus.

Meanwhile, an older white couple waved at the security agent and strolled straight past the booth unimpeded. They weren't asked to show identification, nor were they questioned or prevented from entering the scenic school grounds. This prompted the coach to request a conversation with the security officer's supervisor, who, when he arrived on the scene, eventually gave the go-ahead signal. At which point, the first security guard escorted the team to the locker room. Being escorted is also not customary. But then, in a matter of minutes, he loudly admonished the boys: "Time to go! You've had enough time in there!"

These student-athletes, who had only just returned from competing on behalf of the school, were forced to leave campus. The fact that none of their parents had yet arrived to retrieve them didn't matter. Basically, they were told to get the fuck out. To make matters worse, it was late evening, the night sky was already a starless black, and the school is located in a predominantly white neighborhood several miles from our home. Can you imagine the alarm I felt when my wife and I arrived only to learn that our son wasn't there, that a school official had kicked him off campus, and that he had had to find an open, public establishment that would allow him to hang out until we showed up? Visions of Trayvon Martin and Tamir Rice, the twelve-year-old black boy shot and killed by a white Cleveland police officer, ran through my head. I was horrified. I wanted to know who in the hell was responsible for putting my child at such risk.

I immediately emailed school officials to request a meeting. More than two weeks passed before one was held. Being put off for weeks pissed me off, especially given the nature of my concern and the fact that we had previously met with the school's leadership about the discriminatory practice of their security force.

Even so, I tried to remain optimistic. I told myself that they must be using the intervening time to implement corrective actions. I had hoped the officials would take this issue seriously. I had hoped they would understand how the discrimination carried out by their security personnel could traumatize Malakai and result in cascading negative consequences for his psychological development and functioning. I had hoped they would understand that their inaction *was* the problem, that their inaction could insidiously shape my son's perception of himself and his place in the world, and that their inaction was not so subtly encouraging him to adopt a self-protective, hypervigilant, overly suspicious, even paranoid, and definitely subservient posture. I had hoped they would understand that this kind of race-based, dehumanizing treatment, especially during adolescence, produces significant harmful effects on the mental health of black boys and that these effects continue well into adulthood decades later.[15] And this is to say nothing of the whole host of other possible pathological conditions, including cardiovascular disease, that occur at higher rates among those subjected to racial discrimination.[16]

Nope, I was dead wrong, and this was clear from the meeting's outset. The head of school and her team neither took responsibility nor presented a concrete corrective plan. In fact, they disingenuously asked us, Malakai included, to help craft a school-security policy that would not discriminate against black boys. By this point in my life, I could immediately recognize this frequently employed deflection strategy. The school officials concocted this phony request in their attempt to feign a willingness to take action without actually doing anything. It was insulting. Neither Malakai, nor Robin, nor I have any expertise on security matters, so asking us to solve this problem for them was clearly inappropriate. What's more, even though the head of school knows that I'm a professor at a major university, she has never sought my assistance on curriculum content or development. Seeking my views on issues related to my actual profession is considerably more appropriate than seeking them on questions of security.

So, with this as a backdrop, the pearl-clutching horror expressed by school officials in response to my child's being called a "nigger" was simultaneously unmoving and crazy-making. Here's why: the harm caused to Malakai—and to other black boys—from being called a nigger by an ignorant student pales in comparison with the myriad harms caused from repeated acts of discrimination sanctioned by the school and perpetrated by adults.

These thoughts weighed heavily on my mind on the night of February 21. I had just taken Kenya, our dog of fourteen years, outside to do her biological business for the final time. I was anticipating Malakai's call to scoop him up from school following yet another track meet. Added to my stress was the usual gloom this particular date brings each year. In 1965, Malcolm X—one of my heroes because of his public courage and integrity—was assassinated and, in 2011, Bob Schuster, a beloved friend and former director of NIDA, died on the same day.

Now we were minutes away from euthanizing Kenya. Her health had gravely and irreversibly declined: she had lost her sight, could barely hear, was intermittently confused, had stopped eating, and was now nearly immobile. Despite this, I agonized over my decision to kill her. Intellectually, I knew it was the right thing to do. Emotionally, I felt just the opposite. "Am I doing the right thing?" I wondered silently. The grief-stricken expression on my son Damon's face as he cuddled Kenya one last time filled me with doubt and despair. I struggled to fight back my own tears.

"Which drugs are you using?" I asked the husky veterinarian as she squatted uncomfortably on our kitchen floor. The list of drugs she rattled off were familiar: xylazine, acepromazine, and pentobarbital. "Acepromazine," I asked, "that's a major tranquilizer as well as antihistamine, right?" She gave a response that suggested she wasn't sure. But it really didn't matter because I had only asked the question to distract myself, to suppress my overwhelming emotions, and to dam off the inevitable flood of tears threatening to publicly expose my anguish. Over the years, I have learned that focusing on the details of how drugs work is an effective strategy for dodging and concealing my emotions. So, in that moment, my mind busied itself with speculations about the chemical structure of acepromazine in relation to other neuroleptics and antihistamines, such as chlorpromazine and promethazine. I wondered how selective each was for dopamine versus histamine receptors.

After three quick injections, it was over. Kenya relaxed into a heavy sleep; she appeared at peace. Her breaths slowed and deepened. Then she was dead; her face, mask-like, held an unfamiliar expression. It was lifeless, disconcerting. It still haunts me.

Almost immediately after Kenya had been put to death, Malakai phoned, and I headed out to get him. On the way, I worked out in my head what I'd say to him about Kenya's death and about how she taught me to love. I wanted him to know that it was OK to cry, that it was OK for him to express his own anguish about having lost our dear dog. I wanted him to know that even I was heartbroken and suffering tremendously. But when he got in the car, I broke the news to him with few words and even less emotion. He responded in kind. We rode home mostly in silence, separately suffering in parallel.

I felt for him but didn't know what to say. I felt for him because he had learned, at an early age, as I had, to hide such feelings in order to protect himself. He had learned to compartmentalize in order to appear less vulnerable. I grew up in a world in which boys didn't cry, or, at least, they didn't let anyone see them do so. Crying—or a similar emotional expression—was interpreted as weak, soft. Naturally, I taught my sons this, too.

But I wasn't thinking about their human need to emote in response to situations such as the death of a loved one. Instead, I was exclusively focused on inoculating them from being ravaged by American anti–black boy racism. I didn't want my sons to be perceived as weak because I knew, if they were, they'd be finished. American racism has a way of detecting and devouring weak black boys like a gray wolf pack in pursuit of a wounded deer.

Still, in that moment, I felt for Malakai. I felt for him because he was growing up in a world where black boys still had to conceal their emotions. I felt for him because he was growing up in a world that insisted on treating him like a nigger, just as it had treated his father before him and my father before me and his father before him.

By the time we arrived home, the vet had taken Kenya's body away. We, as a family, went about storing her belongings and preparing for the next day as if it were just another day. It wasn't, of course. And I knew eventually I would have to circle back with Malakai and check in to make sure he was OK; I mean *really* OK. I just couldn't do it in that moment.

First, I had to do my own head work. This work is beyond the expertise of typical U.S. therapists, who are rarely trained in matters of race, especially as it relates to the impact of racism on mental-health functioning and on parenting. So I cleared my schedule for the next few days and shut out the world, and Robin and I took a respite facilitated by multiple doses of my favorite novel psychoactive substance, 6-APB. It was wonderfully cathartic and healing. 6-APB is the quintessential psychoactive ingredient for nourishing the dispirited soul. It helped me to gain new insight into past struggles and into the ones staring me in the face. I came away from this, as well as other 6-APB experiences, feeling replenished and magnanimous. I phoned my friend David Nichols, the guy who first synthesized 6-APB, and wistfully expressed something like, "If only more people could have similar experiences, perhaps we'd treat each other, even strangers and rivals, more humanely." Dave responded, "Yeah, I get that a lot."

7

Cannabis: Sprouting the Seeds of Freedom

I do smoke, but I don't go through all this trouble
just because I want to make my drug of choice legal.
It's about personal freedom.

Bob Marley

I t's me again Jah," Luciano's mesmeric baritone plea filled my headphones, "I pray my soul you'll keep." His voice and lyrics were complemented by the unusually quiet, beautiful New York morning. I had just stepped off the subway onto Columbia's Morningside Heights campus and strolled past Low Memorial Library. With the chapel in my field of vision, I couldn't help thinking back to my youth, when my mom made me attend church regularly. I hated it. It was an October Sunday in 2016.

In an attempt to diminish the anxiety triggered by this memory, I quickly juxtaposed pleasant thoughts of my free and easy weed-smoking evenings, immersed in my music. The music came alive. I heard every instrument, including those that were silent in my sober state. Each one requested, but did not demand, my undivided attention. Marijuana narrowed my focus to the engaged activity by minimizing cognitive intrusions. It helped me to place in abeyance my perpetual mental war preparation in the face of the perils of being black in the United States so that I could be in the moment and enjoy it.

I was headed to my office to complete unfinished work and prepare for the demands of the upcoming week. Since I had become department head three months earlier, I was rarely left alone to simply think, much less to work uninterruptedly on projects that mattered to me. So my Sundays alone in my office had become a fiercely guarded refuge.

Comforted and strengthened by Luciano's divine voice, I was already savoring my Sunday solitude even before arriving at my office. Upon entering the building, though, I was greeted by an unfamiliar fortysomething white guy wearing a gray suit, a crisp white shirt, and studious black specs. His attire was out of place for Schermerhorn even on a workday, let alone on the weekend. *Strange*, I thought. But I tried to keep it moving by avoiding eye contact. I failed. Eye contact was made, which I quickly broke only to get hit with his infectious smile. The broad grin on the stranger's face suggested that he knew and was expecting me.

My Sunday sanctuary was now threatened by this well-dressed, warmly smiling gentleman, who stood between the path leading to my office and me. To make matters worse, Luciano was no longer singing in my head. His voice had been replaced by my own, which silently repeated one simple word: "Fuck!" The phony smile on my face concealed the contents of the mind of a man who desperately wanted to be left alone.

"Hi, Dr. Hart," the rich voice, tinged with a polite southern accent, belted out enthusiastically, "I'm Mike Schneider." As he extended his hand for the traditional mainstream physical greeting, my mind raced trying to place him. "Who the fuck is this guy?" I thought to myself, while continuing to sport a fake-ass smile.

The quintessential absentminded professor, I had forgotten we'd arranged to meet at this time. I would later remember that we chose this particular Sunday because this fellow was in town to attend an art show over the weekend.

A few months earlier, Judge Schneider had sent me an unprompted email explaining that he was in charge of a juvenile drug court in Harris County, Texas, the third most populous county in the United States. Over the course of the six years he had run his drug court, he had grown concerned that too many young people were being diagnosed with substance-use disorders simply because they possessed or had used an illegal drug. Overwhelmingly, cannabis use was the triggering factor for the court's involvement. Numerous teens were being ordered to complete drug-addiction treatment programs for merely having used cannabis, even those whose drug use did not rise to the level of addiction.

Appropriately, Judge Schneider worried that some drug courts were overtreating or mistreating many of the people they were trying to help. He wanted my input on what his drug-court team was doing and what they should be learning in order to better serve those coming before him.

By the end of my first meeting with this affable judge, the discussion had covered a range of topics, from DJ Screw's untimely death to how our society overcriminalizes cannabis use by black and Latino teens in Harris County. I learned that he'd been on the bench for more than seventeen years and that, relative to other judges with whom I had interacted, he was uncharacteristically compassionate toward drug users and open-minded about drug use. Judge Schneider's innocent warmth and openness, combined with his intellectual curiosity and genuine interest in improving the lives of those he served, were so impressive that I agreed to visit his Houston courtroom and meet with his team.

But the thing that I remember most about our first meeting was the discussion about marijuana and the changing attitudes about its use and regulation. We speculated about why there hadn't been appreciable movement toward legalizing recreational cannabis in the southern portion of the United States. At the start of 2016, weed was legal for adult use in four states: Alaska, Colorado, Oregon, and Washington. By the end of that year, four more states had legalized the drug: California, Maine, Massachusetts, and Nevada. None of these states had a black population as high as the national average of 12 percent.

By contrast, the proportion of black citizens living in many southern states is larger than the national average, and cops in these regions routinely cite the smell of cannabis as justification for stopping, searching, or detaining black people. Judge Schneider speculated that the law-enforcement community and their supporters would vigorously oppose any legislation seeking to liberalize cannabis laws because they were acutely aware that claiming to detect the weed's odor is one of the easiest ways for officers to establish probable cause, and judges almost never question the testimony of cops.

What's worse, there have been countless cases during which officers cited the fictitious dangers posed by cannabis to justify their deadly actions. On July 6, 2016, in St. Anthony, Minnesota, officer Jeronimo Yanez shot and killed Philando Castile, a defenseless black motorist, as his girlfriend and young daughter watched helplessly.

Castile informed the officer that he had a firearm on him, for which he had a permit. But within a matter of seconds, Yanez had fired seven slugs into Castile for no apparent reason. The smell of weed, Yanez claimed, constituted an apparent imminent danger. He was acquitted of manslaughter.

In typical fashion, both gun-rights and marijuana advocates were practically silent about this injustice. Their deafening silence has become shamefully too common when the victim is a black male.

Of course, this wasn't the first time, nor will it be the last time, that a police officer has cited the "cannabis makes black people homicidal" defense to justify using deadly force. Michael Brown, of Ferguson, Missouri, in 2014,[1] and Keith Lamont Scott, of Charlotte, North Carolina, in 2016, were both killed by police who used some version of this bogus defense. Neither officer was charged.[2]

Ramarley Graham (New York, 2012), Rumain Brisbon (Phoenix, 2014), and Sandra Bland (Prairie View, Texas, 2015) also had their lives cut short as a result of an interaction with law enforcement initiated under the pretense of cannabis-use suspicion.

But none of these travesties of justice hit home more acutely than the killing of seventeen-year-old Trayvon Martin. On the evening of February 26, 2012, he walked out of a local 7-Eleven with the items he'd just bought—a can of AriZona watermelon fruit-juice cocktail and a bag of Skittles—and headed to his father's home. But before the unarmed teen could safely arrive, he was scoped like game, stalked as if in the wild, and fatally shot by the neighborhood vigilante, George Zimmerman.

I still remember my initial feelings of shock as if it were yesterday. I was incredulous when I learned the details of this monstrous and brutal act. The twenty-eight-year-old Zimmerman, who identifies as white, phoned the local police nonemergency number after merely spotting Trayvon en route to his father's place and claimed the teen looked like "he's on drugs." For no apparent reason, Zimmerman then chased the youngster, ignoring the dispatcher's directive to the contrary. "We don't need you to do that," the dispatcher admonished Zimmerman. Minutes later, he had drawn a 9 mm semiautomatic pistol and killed a child in cold blood.

I thought about Trayvon's parents, and I wept. He could've been my child. I thought about my own son Damon, who was the same age as Trayvon. This realization filled me with dread. I wept even more. I was profoundly sad. What is worse is that I felt powerless to protect children who look like me.

And then, to add insult to injury, Zimmerman's lawyers blamed Trayvon for his own death. They argued that the teen was aggressive and paranoid from smoking marijuana and that led him to attack their client. The Zimmerman team reverted to the familiar and tired marijuana-crazed Negro script, illustrating the enduring persuasive power of this myth. Dragging a deceased child's reputation through the mud in the process was cruel and deplorable.

This pissed me off. I got a copy of Trayvon's toxicology report, carefully examined it, and wrote an op-ed pointing out that the notion that weed causes violence is an inexcusable fallacy.[3]

By that time in my career, I had given literally thousands of doses of marijuana and had completed multiple studies assessing the neurophysiological, psychological, and behavioral effects of the drug.[4] I had never seen a research participant become violent or aggressive while under the influence of cannabis, as Zimmerman's team was claiming when describing Trayvon's actions.

The main effects of smoking marijuana are contentment, relaxation, sedation, euphoria, and increased hunger, all peaking within five to fifteen minutes after smoking and lasting for about two hours.

True, very high THC concentrations—far beyond Trayvon's levels—can cause mild paranoia and visual and auditory distortions, but even these effects are rare and usually seen only in very inexperienced users.

The toxicology exam, which was conducted the morning after Trayvon was killed, found a mere 1.5 nanograms of THC per milliliter of blood in his body. This finding strongly suggests he had not ingested marijuana for at least twenty-four hours. This level is also far below the THC levels that I have found necessary, in my experimental research on dozens of subjects, to induce intoxication: between 40 and 400 ng/mL. In fact, his THC levels were significantly lower than the sober, baseline levels of about 14 ng/mL of many of my study participants who are daily users. Trayvon could not have been intoxicated with marijuana at the time of the shooting, as claimed by Zimmerman; the amount of THC found in Trayvon's system was too low for it to have had any meaningful effect on him.

But it didn't matter. By introducing Trayvon's toxicology results, Zimmerman's defense team created enough of a smokescreen for the jury to find the white neighborhood vigilante not guilty in the killing of a defenseless black boy.

Back in the 1930s, numerous media reports exaggerated the connection between marijuana use by blacks and violent crimes. Some people even claimed that marijuana use was a cause of matricide. These fabrications were used to justify racial discrimination and to facilitate passage of the Marijuana Tax Act in 1937, which essentially banned the drug. During Congressional hearings concerning regulation of weed, Harry J. Anslinger, commissioner of the Federal Bureau of Narcotics, declared, "Marijuana is the most violence-causing drug in the history of mankind."[5]

As we see, the reefer-madness rhetoric of the past has not evaporated; it has evolved and reinvented itself. In his recent book *Tell Your Children: The Truth about Marijuana, Mental Illness, and Violence*, Alex Berenson writes, "Marijuana causes paranoia and psychosis."[6] This opinion does not align with the evidence from science. I think it's important for you to understand how someone might arrive at this interpretation, because with each new generation, the myth of reefer madness is revamped and disguised as empirical evidence rather than as what it is: misinformed rhetoric.

For starters, knowing something about how studies in this area are conducted should reduce your susceptibility to accepting wild claims. Typically, a few thousand adults are separated into groups based on their reported current or prior use of marijuana: marijuana users in one group and nonusers in the other. Then, researchers see if the groups differ on measures of psychosis.

The first questions you should ask are, What is psychosis? and How is it determined or diagnosed? From a clinical perspective, psychosis is a mental disorder involving a loss of contact with reality and is characterized by hallucinations, irrational beliefs, and disorganization of speech and behaviors. Experts typically think of psychosis in association with schizophrenia, but it can also be present in other disorders. For someone to be diagnosed with psychosis, that person must be evaluated by a psychiatrist or psychologist. This evaluation can be rather involved and time-consuming.

In most such studies, research participants are not assessed for a psychotic disorder. Such assessments are too impractical. Instead, participants complete questionnaires, containing about twenty items, that probe psychotic *symptoms*. One consistent finding is that participants in the marijuana group are more likely to have experienced at least one psychotic symptom. Of course, this finding doesn't mean that these individuals have a psychotic disorder. Much more information is needed before that determination can be made. But unfortunately, this crucial caveat is poorly understood. That's one reason the public is inundated with sensational and misleading headlines, such as EVEN INFREQUENT USE OF MARIJUANA INCREASES RISK OF PSYCHOSIS BY 40 PERCENT.[7]

To be clear, it is possible to experience psychotic *symptoms* without meeting criteria for a psychotic *disorder*. Many people, including me, have at some point in their lives experienced at least one of these symptoms. But considerably fewer of us have ever met the requirements for a disorder. This point becomes clearer once you've checked out a few actual symptoms listed on typical psychotic questionnaires. "I hear voices that others do not" and "I sometimes feel uncomfortable in public" are two such items. It's not difficult to see how the *former* might be clinically meaningful, whereas it seems a bit of a stretch to suggest the *latter* is uniquely related to psychosis. Also, it's important to keep in mind that symptoms contained on these questionnaires may be experienced only for a brief period and do not necessarily indicate a permanent state.

A related point deals with how causation is determined. It is true that people diagnosed with psychosis are more likely to report current or prior use of marijuana than are people without psychosis. The simple, but uncritical, conclusion to draw from this is that marijuana use causes psychosis. But this interpretation ignores evidence showing an even stronger link between tobacco use and psychosis,[8] to say nothing about the demonstrated associations between psychosis and the use of stimulants.[9] This interpretation also ignores data from multiple studies showing that cat ownership in childhood is significantly more common in families in which the child later was diagnosed with a psychiatric disorder such as schizophrenia.[10]

Do all these things "cause" psychosis, or is there another, more likely answer? One of the most fundamental lessons I try to teach my students is that a correlation or link between factors does not necessarily mean that one factor is the cause of another. For example, there is a strong correlation between the number of umbrellas put up and the amount of precipitation, but it would be foolish to conclude that putting up umbrellas causes rainfall.

In 2016, Charles Ksir and I conducted an extensive review of the literature and concluded that those individuals who are susceptible to developing psychosis (which usually does not appear until around the age of twenty) are also susceptible to other forms of problem behavior, including poor school performance, lying, stealing, and early and heavy use of various substances, including marijuana.[11] Many of these behaviors appear earlier in development, but the fact that one thing occurs before another is not proof of causation. (One of the standard logical fallacies taught in logic classes is, "After this, therefore because of this.") It is also worth noting that tenfold increases in marijuana use in the United Kingdom from the 1970s to the 2000s were not associated with increased rates of psychosis over this same period—further evidence that changes in cannabis use in the general population are unlikely to contribute to changes in psychosis.

Decade after decade, the public has been gaslighted regarding the real effects of cannabis. A stream of blatant lies has been promulgated to justify marijuana prohibition. But as Martin Luther King Jr. once remarked, "No lie can live forever." The fact that weed is America's most widely used banned substance—twenty-seven million U.S. citizens use it monthly[12]—makes it difficult to peddle distorted claims about the drug and be believed. Too many people now have marijuana-use experiences that conflict with the cannabis gaslighting engaged in by public officials. In recent years, it has become infinitely harder to fool the public about the actual effects produced by marijuana, although some police officers who kill black people still manage to do it. It also can't be ignored that marijuana use, compared with other drug use, is less stigmatized. Consequently, weed smokers are much more likely than users of other drugs are to be out of the closet, and they are quick to call gaslighters out on their bullshit publicly. It is my hope that weed smokers will also begin to speak out in defense of brutalized black people who have been accused of fictitious acts as a result of marijuana use.

There is now a flourishing movement to liberalize policies that regulate weed in the United States and beyond. You may recall that marijuana is still listed on Schedule I under the federal Controlled Substances Act. This designation denotes that cannabis has "no acceptable medical use in treatment" and that it is banned. Federal law still forbids the use of marijuana as a medical treatment. In practice, placing aside the fact that since 1976 the U.S. federal government has supplied marijuana to a select group of patients, citizens throughout the country have voted repeatedly to legalize medical marijuana at the state level. Since 2010, for example, the number of states that allow patients to take the drug for specific medical conditions has jumped from sixteen to thirty-three—a number that is expected to climb steadily with each successive election season.

If federal marijuana law is the theory and state-level ballot initiatives are the practice, then Albert Einstein's words are fitting: "In theory, theory and practice are the same. In practice, they are not." There are gaping inconsistencies between what is stated in the federal marijuana law and the clinical realities practiced in an increasing number of states.

This regulatory duplicity hasn't been lost on Congress, at least judging from the general tenor of a congressional hearing on cannabis I participated in a few years ago. Typically, such hearings can be summed up as exchanges of misinformation, dominated by antidrug jingoism spouted by both Democrats and Republicans. These time-honored, choreographed productions are usually impervious to data.

Not this one. These congressional committee members were not just supportive of removing marijuana law restrictions; some were also downright hostile toward witnesses who engaged in marijuana gaslighting and fearmongering.

In response to NIDA director Dr. Nora Volkow's biased appraisal of marijuana-related effects, Virginia Democratic representative Gerald Connolly's tone shifted from receptive to confrontational. In his defense, it didn't take much effort to recognize Dr. Volkow's inappropriate partiality. Her testimony sounded more like an antimarijuana scare campaign from an earlier time—focusing narrowly on potential toxic effects—than like a reasoned and objective presentation of the current scientific understanding of the drug and its effects.

"Dr. Volkow, your testimony seems to completely disregard lots of other data," Connolly accused her.[13] And he was right. Omitted from her remarks were any mention of scientific findings that demonstrate, for example, that marijuana improves mood, sleep, and appetite, among a host of other favorable outcomes. That she so blatantly neglected to acknowledge any potential beneficial effect derived from cannabis prompted an exasperated Connolly to state that the actions of NIDA have "impeded the ability to have legitimate research that could benefit human health."

It seems only a matter of time before Congress weighs in on whether marijuana should be legalized nationwide.

The free-the-weed movement is not limited to medical marijuana. A growing number of prosecutors from states where recreational use of the drug is still prohibited say they will no longer prosecute marijuana possession cases. This move essentially legalizes weed consumption, but not sales, in places such as Baltimore, Brooklyn, Chicago, Manhattan, and Philadelphia. In some jurisdictions, previous marijuana possession convictions are even being cleared off the books.

Well, it's about fucking time! Officials in these cities have known for many years that their police departments were carrying out arrests for marijuana possession in a racially discriminatory manner. Even worse, this type of racism continued despite the fact that possession of the drug had been decriminalized in each of these cities. Take Baltimore, for example, where blacks represent about 60 percent of the population and about as many weed smokers. There decriminalization took effect in October 2014. Yet between 2015 and 2017, Baltimore police arrested 1,514 individuals for weed possession; *1,450* were black.[14] That's 96 percent. That's racial discrimination. That's shameful.

Undoubtedly, the glaring spotlight exposing racism helped these prosecutors see beyond the smokescreen that exaggerates the dangers to public safety posed by marijuana users. Baltimore's chief law-enforcement officer Marilyn Mosby justified her decision to no longer prosecute marijuana-possession cases by stating, "Prosecuting these cases has no public safety value, disproportionately impacts communities of color and erodes public trust, and is a costly and counterproductive use of limited resources."[15] I agree.

These prosecutorial decisions come at a time when nationwide support for legalizing cannabis is at an all-time high—no pun intended. As Figure 4 shows, in 2018, a whopping 66 percent of Americans said weed should be legalized, reflecting a steady increase over the past three decades.[16] Correspondingly, as of early 2020, eleven states and Washington, D.C., now allow recreational marijuana use by adults (Table 2). This number will most certainly grow over the next few years because several governors recently made support for legalizing cannabis a key issue in their successful bids for office. Moreover, Uruguay, in 2013, and Canada, in 2018, became the first two nations to fully legalize weed.

FIGURE 4

Support for Legalizing Marijuana Continues to Edge Up

Do you think the use of marijuana should be made legal, or not?

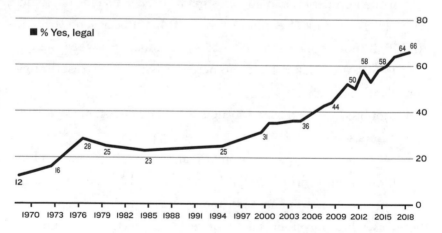

The proportion of Americans who support marijuana
legalization has dramatically increased since 1970.

TABLE 2

State	Year Passed
Alaska	2014
California	2016
Colorado	2012
District of Columbia	2014
Illinois	2019
Maine	2016
Massachusetts	2016
Michigan	2018
Nevada	2016
Oregon	2015
Vermont	2018
Washington	2012

As of January 2020, eleven states and
Washington, D.C., have passed laws legalizing
recreational marijuana.

I frequently think about the changing landscape of marijuana regulation and wonder how we've gotten here. Is there now new scientific data contradicting the stories told in the 1930s that led to the weed ban? Is science the driving force? How do we square the fact that numerous mothers continue to forfeit custody of their children for merely using the drug, especially if the use occurred during pregnancy? Would we tolerate children being removed from their mother just because she drank a glass of wine?

Perhaps all the recent scientific evidence is stimulating marijuana policy changes? That question pops in my head every time I fly in or out of LaGuardia Airport. For it turns out that we've known quite a bit about the effects of marijuana for decades. Around the time the drug was banned, New York City mayor Fiorello LaGuardia commissioned a comprehensive study of its use and effects. Findings from the LaGuardia report were published in 1944,[17] and they were clear but inconsistent with the rhetoric that had led to passage of the Marijuana Tax Act. In short, the conclusions were that individuals "who have been smoking marijuana for a period of years showed no mental or physical deterioration which may be attributed to the drug," and that concerns about catastrophic effects of smoking weed were unfounded. These findings are congruent with data from the hundreds of subsequent studies on marijuana, including my own. I doubt that new information about weed's effects has been the impetus for recent policy changes.

Is the current movement being driven by an increase in weed use? Probably not, because marijuana-use rates were highest in the late 1970s. For example, only about 45 percent of high school seniors report having ever smoked marijuana in 2019,[18] compared to 60 percent of seniors in 1979.[19]

There is little doubt that money has hugely contributed to people's change in attitude about whether marijuana should be legalized. It doesn't take a rocket scientist to see that states can generate millions of dollars each year in new tax revenue as a result of legalizing the drug. For example, the Colorado Department of Revenue reported that marijuana sales in 2018 were just under $1.55 billion. Sales have increased each year since 2014, when the first year of legal pot generated $683.5 million. In terms of tax revenue generated for the state, the amount exceeded $266.5 million in 2018, up from $247 million in 2017. From 2014 to 2018, the state's marijuana taxes generated nearly $1 billion in tax revenue for the citizens of Colorado.[20] Other states are paying attention.

I'VE THOUGHT A lot about all those mothers who lost custody of their children because they used cannabis during pregnancy. I wasn't even aware of the extent of the problem until it was brought to my attention by Lynn Paltrow, executive director of National Advocates for Pregnant Women (NAPW). I was horrified to learn about the many injustices faced by pregnant women simply because they may use drugs. I wanted to help, so I joined the NAPW board of directors.

Can you imagine being told that your child is better off without you merely because you smoked a joint? That's the awful reality staring too many women dead in the face, even in states where the drug is legal. In the eyes of the court, marijuana use by pregnant women is tantamount to child abuse.

Nowhere was this more apparent than at the 2019 Texas Juvenile Law Conference. I rarely attend conferences nowadays, but I went to this one. I went to give a lecture on the myths and realities surrounding cannabis. Judge Mike Schneider and the other conference organizers thought my perspective would be instructive for the attendees. Mike and I had become friends over the past two and a half years, which made it difficult for me to decline the organizers' invitation, but I was also glad for the opportunity to educate family-court practitioners, including judges.

The large ballroom was filled with Texas law-enforcement officials—male, female, Asian, black, Latino, white. These were not my people—cops, prosecutors, judges, probation officers. They made me uncomfortable. But my discomfort had more to do with the lingering psychological impact of hostile law-enforcement encounters I've had in the past.

As part of my talk, I told the audience that, of course, excessive drug use should be discouraged during pregnancy. It doesn't matter whether the drug is alcohol, caffeine, tobacco, or weed. Consumption of any of these substances should be taken in consultation with a health-care professional and should be limited. Similarly, women who eat unhealthy diets while pregnant risk putting their and their fetuses' health in jeopardy. However, I also explained that the negative health consequences of prenatal marijuana exposure for child development are frequently overstated. The exaggerations, in themselves, can harm women and their children by increasing the unwarranted stigma associated with cannabis use by expecting mothers. In the past, this intense stigma has resulted in the removal of children from their mothers and even in the incarceration of mothers.

When I finished speaking, the huge packed room was silent. This crowd wasn't accustomed to hearing anything that conflicted with the message "Drugs are bad, period!" Slowly, a few people spoke up. Their assertions, usually cloaked in the form of questions, made it clear that they believed, "You are either *discouraging* marijuana use or *promoting* it." I tried to help these legal experts see that there are more nuanced ways of thinking about cannabis specifically and drugs in general. The twenty-minute question-and-answer session dragged on for what seemed like forever. I felt as if I had been teleported back to the 1980s, when Nancy Reagan's slogan "Just Say No" passed as high-level drug education.

"You do realize you're in Texas, right?" one guy asked, sporting a self-satisfied smirk on his weathered red face. Not being entirely certain of the intention behind his question, I shot him a mechanical smile that was about to be followed by some Miami-mean shit. But my smile must've shown sufficient appreciation of Texas's special-ness because he immediately moved on to the crux of his issue. He wanted me to know that I was wrong, that prenatal weed use un-equivocally produces cognitive impairments and other abnormali-ties in offspring. He even cited a "scientific expert" who shared his position: Dr. Ira Chasnoff.

A pediatrician by training, Chasnoff may be best known for ex-aggerating the number of fetuses exposed to cocaine prenatally and for wildly overstating the effects of prenatal cocaine exposure on child development. In the 1980s, without the necessary evidence, he and others warned of the extraordinary horrors that awaited so-called crack babies as they aged. Astonishingly, Chasnoff advised parents of cocaine-exposed infants "not to make too much eye con-tact with their babies" because it overwhelmed the child.[21] Not only was this advice unfounded, it also conflicted with the major theories on parent-child bonding. This was reprehensibly irresponsible.

Despite this, in 2017, Chasnoff was at it again. Now focused on prenatal marijuana exposure, he published an editorial in the *American Journal of Obstetrics and Gynecology*.[22] Neither objective nor informative, the piece suffered from misinterpretations of previous research findings to draw an expedient conclusion. For example, Chasnoff stated that "a consistent pattern of deficits" has been ob-served in prenatally marijuana-exposed children. This is simply in-correct.

The totality of the evidence shows that on the overwhelming majority of measures, the cognitive performance of marijuana-exposed children does not differ from that of control subjects. Furthermore, even when there is an observed statistical difference, it is inappropriate to conclude that that difference equates to a deficit, or that it has an impact on the daily functioning of the individual. That is why it is essential to determine whether cognitive scores are within those of the normal population's range.[23] If scientists (as well as non-scientists) are not cognizant of this potential pitfall, they run the risk of inappropriately labeling children as defective, just as was the case during the so-called crack-baby epidemic.

My major concern is not that pregnant women might be advised to avoid using cannabis. They already receive sound advice related to nutrition, environmental hazards, and substance use. Rather, I worry that vociferous misinformed moralists, masquerading as scientists, misrepresent the available data on prenatal cannabis exposure and unjustifiably promote fear. Such reckless behavior has contributed to children being separated from their mothers and placed in foster care. These placements can be far more harmful to children than their mother's cannabis use can.[24] The fact is that many parents who use drugs are good parents, and their children are clearly better off with them.

Unfortunately, this fact is frequently dismissed. For example, a recent proposal calls for pregnant women—or women at risk of becoming pregnant—to be screened for marijuana use but not for their ability to parent.[25] This is preposterous. It uniquely restricts women's civil liberties by exposing them to legal consequences that men do not face.[26] And considering the pervasiveness of racism in drug-law enforcement, black women can expect to bear the brunt of the consequences resulting from this or any other proposed draconian policy.

During a recent visit to a weed store in Denver, I was struck by this thought: *"The public discourse on cannabis too often omits the joy that people seek and experience from the marijuana high."* I watched a continuous stream of people—young, old, female, male, seemingly all law-abiding—flow in and out of the shop. They wore a familiar secret expression: *"I can't wait to hit this . . . to get nice."* Politely and unobtrusively, they all selected their weed, paid for it, and left. I did the same.

Later that evening, I connected with friends and shared the items I had purchased. Marijuana both enhanced the pleasantness of our moods and of the evening, but it also promoted such prosocial behaviors as sharing, openness, and friendship. Considering how emotionally taxing life can be, it is foolish and infantilizing to enact laws that forbid access to this pleasure-producing plant. I'm now firm in my belief that marijuana is a key ingredient to happiness for a great number of people. What kind of person prevents another's responsible pursuit of happiness? Not a very humane one.

8

Psychedelics: We Are One

If the words "life, liberty, and the pursuit of happiness"
don't include the right to experiment with your own
consciousness, then the Declaration of Independence
isn't worth the hemp it was written on.

Terence McKenna

In recent years, psychedelic drugs have become chic. Now, even square, middle-class conventionalists openly boast about their experiences under the influence of these substances, the authenticity of their psychedelic adventures magnified if they take place in an exotic location and are led by a shaman or some other tradition-claiming leader. Psychedelics comprise drugs that include lysergic acid diethylamide, better known as LSD; psilocybin, the active ingredient in magic mushrooms; dimethyltryptamine, also known as DMT, the psychoactive component of the ayahuasca brew; and other substances that are capable of producing profound alterations of perceptions and emotions.

Not long ago, I was exercising in the Columbia gym when a middle-aged white military veteran recognized me and wanted to share his exploits. He didn't refer to these substances as psychedelics; he called them "plant medicines." It was particularly important to him that I know he "didn't get high" and only used the plants to facilitate his "spiritual journey."

"What's wrong with getting high?" I asked with a deadpan look on my face. Like a deer caught in the headlights, the vet froze before babbling a defensive and incoherent thought. Mind you, I knew it had taken him a fair amount of courage to share his story with me, so I didn't want to come across as contemptuous. But that's exactly what I felt: contempt. I wasn't angry with him personally. It's just that I had grown increasingly annoyed with the mental gymnastics that some psychedelic users perform in order to distance themselves from other drug users. The irritation I felt toward a few proponents of psychedelics was now starting to color my view of an entire class of drugs. I knew this wasn't right.

Many people have shared with me their positive life-altering experiences after having used such drugs as ayahuasca, LSD, and psilocybin. Some have said they used them to feel at one with the universe or to sense their connectedness with fellow humans; others have said they used them to explore the meaning of their lives or to experience a kaleidoscope of magnificent colors and images.

Frankly, they seemed to be better people as a result. They are conscientious of others' well-being and express a desire to bring about a more just world. For this, I am grateful. Yet I repeatedly encountered subtle cues of drug exceptionalism, a belief that psychedelics were somehow a superior class of drug. This disturbed me.

Fortunately, a chance meeting at a 2017 Christmas Eve party helped change my thinking about psychedelics. There I met filmmaker Amir

Bar-Lev, who had recently completed his exquisite documentary, *Long Strange Trip*, about the rock band the Grateful Dead. At the time, I knew very little about the Dead and cared even less about their music. I knew the group had a rabid fan base, Dead Heads, who followed them around the country whenever they toured. But without much thought, I had pigeonholed Dead Heads as aging hippies who refused to grow up, and I had neglected to consider the possibility that their use of psychedelics might have been a way for them to explore freedom, a way for them to experience a more meaningful life.

Thus, as I learned more about the band and its journey, it wasn't difficult for me to see why LSD and other psychedelics were key ingredients of the Dead experience. For instance, I doubt people would enjoy the Dead's music as much as they do without the aid of psychedelics.

Amir is a thoughtful, sincere, and unassuming man. He doesn't engage in conversations merely to toot his own horn. He listens attentively and patiently, and makes others want to do the same in his presence. As Amir spoke more about his film, my respect for the Dead, especially for Jerry Garcia, increased. Garcia was viewed by many as the band's leader, but he steadfastly rejected this title up to his death in 1995. He was a firm believer in egalitarian principles and viewed each band member as an equal.

Amir caught me by surprise when he said that my position on adult drug use, specifically that it's an unalienable right, was very much in line with Garcia's view.

"Say what?" I asked incredulously as Amir made his case. I found it difficult to wrap my head around the notion that a psychedelic icon and I might share similar views on drugs. The current popular psychedelic movement seems to be dominated by people who justify their use of these drugs by couching it in medical or spiritual jargon. With some justification: over the past fifteen years or so, a growing number of studies have demonstrated that psychedelics, such as ketamine and psilocybin, produce a range of therapeutically beneficial outcomes, including reducing depressive mood states and provoking spiritually meaningful, personally transformative experiences.[1] The media has warmly embraced these findings and generated a stream of positive buzz in the popular press.

Also, a growing number of popular books and public lectures tout the benefits of psychedelic use. Michael Pollan, in *How to Change Your Mind*, persuasively makes the case that psychedelics, such as LSD and psilocybin, can be personally transformative.[2] A growing amount of scientific data backs him up. In her book *A Really Good Day*, Ayelet Waldman chronicles her monthlong experience of taking small subperceptual doses of LSD to treat her mood disorder.[3] Waldman, like others, was inspired to use "microdoses" of LSD after having read James Fadiman's *The Psychedelic Explorer's Guide*.[4] Despite the fact that there is virtually no solid evidence supporting microdoses of psychedelics to treat ailments or improve performance—the research hasn't been done—microdosing has become the latest fad. Together, these developments have contributed to the removal of the stigma associated with using these substances, so long as the reason for use is not to get high. If your aim is to seek relief from an emotional or physical ailment or to achieve spiritual transcendence or to find your god, cool. But if you merely want to have a good time, not cool.

This arbitrary distinction makes no sense. Oftentimes the alleviation of pain, whether it's psychologically or anatomically based, contributes to feelings of intense well-being and happiness, that is, to a sense of having "a good time." Disentangling these deeply personal and idiosyncratic constructs is a difficult, if not an impossible, task. Similarly, the sacred experiences that positively affect one's self-perception, worldview, goals, and ability to transcend one's difficulties are hard to separate from one's feelings of pleasure or happiness. What's more, psychedelic drugs aren't unique in their ability to produce these responses. I certainly have experienced all these effects after taking heroin or cocaine or MDMA or any number of the drugs that I discuss in this book.

Heroin and cocaine aren't classified as psychedelics; however, MDMA, which is, in fact, structurally an amphetamine, is often classified as a psychedelic. Following a single administration of a large dose (> 250 mg), it's absolutely possible to experience transient, prominent visual perceptual changes sometimes referred to as trails—a series of discrete images following moving objects. But most people who take MDMA aren't seeking visual alterations and will never experience such effects because the typical doses used range from about 75 to 125 mg. Within this dose range, the magnanimous, euphoric, and empathy-enhancing effects that most of us seek from MDMA are far more likely to occur. Thus, if it were up to me—and it's not—I wouldn't label MDMA as a psychedelic. It's an amphetamine, period.

Nonetheless, the discussion raises multiple broader issues about why specific drugs are categorized as psychedelics and others are not. Why have some psychedelics but not others managed to shed ridiculous stereotypes about their effects and users and thereby increase their popular acceptance and respectability? Methamphetamine, a chemical cousin of MDMA, is never referred to as a psychedelic. Why not? At large enough doses, it, too, can produce visual hallucinations. Plus, it has contributed to some of my most transcendent moments and has helped me solidify a few of my most important relationships. The majority of my methamphetamine experiences rival those I've had on MDMA and were a lot more meaningful than those I've had under the influence of so-called psychedelics.

Admittedly, thus far, I have taken only a few psychedelics: 4-acetoxy-DMT, 2-CB, ketamine, and psilocybin. In addition, because of my limited experiences with these substances, I have always taken doses on the low end of the spectrum. As I noted in Chapter 3, the amount of drug taken is one of the most crucial factors in determining the resulting effects. It's possible that I might have found psychedelics more preferable had I pushed the dose. Relatedly, the setting under which drug use occurs can exert powerful influences on how the individual experiences the effects. Many people take psychedelics in the presence of a guide or a shaman, someone who ostensibly serves as a safety monitor as well as an experience interpreter. Some people find this comforting; I find it creepy and have never done so myself. For these reasons, perhaps it's not fair to draw a comparison between my methamphetamine and my psychedelic experiences.

Regardless, the point I'm trying to make is that the way drugs are categorized is often discretionary and dependent upon whomever is doing the classifying. Most people, especially well-educated professionals, classify drugs in the manner that best serves their own purposes. For example, a drug such as MDMA is categorized as a psychedelic by respectable, middle-class white folks because they use and enjoy it. Methamphetamine, however, is not included among the psychedelics because middle-class drug elitists revile it. The perception, in the words of former Oklahoma governor Frank Keating, is that it's "a white-trash drug."[5] Can you imagine if methamphetamine was labeled as a psychedelic? Respectable people would

balk at this notion, fearing that the drug would ruin the rehabili-
tated reputation of their beloved psychedelics. By extension, no
doubt, psychedelic users would worry that their own good names
would be besmirched, too.

However, psychonauts—drug elitists who use psychedelics to
explore altered states of consciousness—can't readily discard phencyclidine, a.k.a. PCP or angel dust, because it has long been established as a psychedelic. You should also know, by the way, that the
term *psychonaut* in itself is another attempt to dissociate middle-
class psychedelic users from users of drugs such as crack and heroin,
who are disapprovingly called "crackheads" or "dope fiends."

In the 1950s, Parke, Davis & Company sought to develop PCP as
an intravenous anesthetic. It was shown to be both safe and effective
in patients. But the medication also produced in some people lingering depersonalization—a feeling of observing oneself from outside
one's body. This effect prompted concern and a more careful study
of the full range of behavioral, neurological, and physiological effects
produced by the drug. In one of the early studies, a large group of psychiatric residents and medical students were administered PCP intravenously at a dose level of 0.1 mg/kg, which is similar to doses used
medically and recreationally.[6] Consistent findings were distortions
of body image and, again, depersonalization. Several research participants reported experiencing pleasurable effects, such as having dream-
like reveries; some displayed disorganized thinking. Not one became
violent. Similar results have been replicated in multiple studies.

We now know that PCP produces many of its effects by selectively
blocking a subtype of glutamate receptors called the N-methyl-D-
aspartate (NMDA) receptor. Glutamate, like dopamine and serotonin, is one of the brain's many neurotransmitters. PCP is a selective

NMDA receptor antagonist, or blocker. Ketamine—developed by altering PCP—is structurally similar to its parent compound (see Figure 5). The drug is also an NMDA receptor antagonist, but it's not as selective as PCP. For instance, ketamine's effects don't last as long as those of PCP, which decreases the likelihood of unwanted side effects. Perhaps this is one of the reasons that ketamine has essentially supplanted PCP in medicine. Indeed, some of the most exciting recent findings from psychiatric research have been obtained using ketamine in the treatment of depression. Its therapeutic effects are observed within twenty-four hours, which is markedly faster than the seven to fourteen days it usually takes before the onset of beneficial results produced by traditional antidepressant medications like escitalopram, fluoxetine, or venlafaxine.

FIGURE 5

Phencyclidine   Ketamine

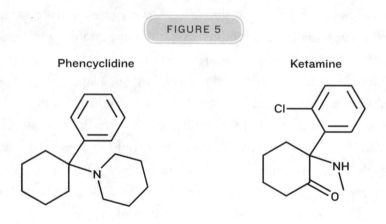

Chemical structure of phencyclidine (a.k.a. PCP: left)
and ketamine (right)

Another factor that has practically eliminated the use of PCP in medicine has to do with claims that illicit use of the drug produces extraordinary violence in users. At some point in the 1970s, multiple media reports emerged making this allegation. Police narratives further cemented the supposed connection between PCP and violence. Perhaps you've heard the story that describes a PCP user who became uncontrollably violent, developed superhuman strength, and was impervious to pain after taking the drug? He had to be shot at least twenty-eight times in order for the police to restrain him—or so the story goes, anyway. Sound familiar? Remember the superhuman Negro cocaine fiend?

The fact is that there is no evidence the events in this story ever happened. It's an urban legend. But that doesn't seem to matter, because the story continues to live on and is retold repeatedly. This legend and others have contributed to the misconception that tremendous force is required when apprehending a suspected PCP user. In 1988, my late friend Dr. John Morgan and his colleagues were suspicious of unsubstantiated claims regarding PCP-induced violence, so they reviewed the clinical literature on the topic and published their findings in a peer-reviewed scientific journal.[7] After carefully assessing nearly one hundred cases in which PCP was said to have caused people to engage in violent acts, the researchers found no connection between PCP and violence. They concluded that popular assumptions claiming PCP uniquely causes violence among its users are simply not warranted.

With solid scientific evidence, you might think the myth of the violent PCP user would eventually die. But that's not how drug myths work. They don't die, they get revitalized with each successive generation. The Rodney King incident is a case in point. In March 1991, four Los Angeles police officers—all white—were caught on camera savagely beating King—who was black—after they had stopped him for a traffic violation. The police beat King so badly that he sustained multiple serious injuries, including skull fractures, broken bones and teeth, and permanent brain damage. During the trial, police said that they used such great force because they thought King was "dusted"—under the influence of PCP. He wasn't. His toxicology report revealed that he had only consumed alcohol.

Despite all this, the four officers were acquitted on April 29, 1992. Los Angeles erupted in protests and civil disobedience that lasted five days. A year later, two of the officers—Stacey Koon and Laurence Powell—were convicted in federal court of violating King's civil rights and sentenced to two and a half years in prison. The events reignited the same national conversation about racism and police brutality that we have had for more than fifty years. Yet, as usual, the discourse was devoid of any interrogation of the mythical "PCP-induced violent criminal" as a credible police defense for the use of excessive force. In many black communities, this omission reverberates even today.

PCP FOUND IN BODY OF TEEN SHOT 16 TIMES BY CHICAGO COP. That was the headline of an article published in the *Chicago Tribune* on April 15, 2015.[8] My heart sank from merely reading the title, even though I had grown accustomed to police claiming that deadly force was necessary because the victim had PCP in his body. It only got worse as I read the entire article: "A knife-wielding teen had PCP in his system . . . it can cause its user to become aggressive and combative." Here we go again.

I remembered having read in passing about the killing of seventeen-year-old Laquan McDonald—who was black—by a Chicago cop back in October 2014. The officer's identity was withheld from the public for several months. But at the scene of the shooting, Pat Camden, police union spokesman, quickly took control of the narrative. He told the media that McDonald "walked up to a car and stabbed the tire of the car and kept walking." When officers ordered McDonald to drop the knife, according to Camden, rather than complying, the teen supposedly lunged at police, prompting one officer to fire his gun. Reporting for the *Chicago Tribune*, Quinn Ford wrote, "McDonald was shot in the chest and . . . pronounced dead" shortly after.[9] The implication was that the officer had fired only one shot while defending himself and other police. The story seemed familiar and a matter of fact. True to form, it was peddled to the public as the official version of the events that transpired on the night of October 20, 2014, when young McDonald was shot dead.

It turns out, the police lied, and the media did a piss-poor job of journalism. The public might not have known this and other crucial details if it were not for the exceptional investigative reporting of Jamie Kalven. Kalven obtained a copy of McDonald's autopsy report through a Freedom of Information Act request. And on February 10, 2015, more than three months after the killing, he published an article in *Slate*.[10] It detailed what Kalven had learned, including multiple pieces of information that impeached—or at least seriously called into question—the story told by the police and the mainstream media. Chief among them were that McDonald had been shot sixteen times, not once, as had been implied, and that the shooting was captured on a dashboard camera in one of the squad cars. Kalven called for the police to release the video.

City officials, including mayor Rahm Emanuel and police top brass, refused to make the video available to the public. They did, however, in collaboration with the media, employ the same tired script that had worked so well in previous incidents involving police misconduct: drag McDonald's reputation through the mud and blame a drug for his alleged destructive behavior. Thus, findings from his toxicology report were released. PCP was found in his system, or so it was widely reported. Unsurprisingly, these accounts emphasized that the drug could make the users aggressive and violent. Not one of the reports mentioned that false positive toxicology screens for PCP are common for several prescription and over-the-counter drugs, including tramadol, venlafaxine, alprazolam, clonazepam, carvedilol, dextromethorphan, and diphenhydramine. Not one mentioned scientific evidence indicating that the link between PCP and violence is not justified.

But then, on November 24, 2015, thirteen months after the incident, a judge ordered city officials to release the dashcam video of the shooting. Its contents were horrifying and utterly contradicted the police account. McDonald was walking away from police when officer Jason Van Dyke—who is white—quickly moved toward the teen, opening fire while standing only about fifteen feet away. Van Dyke, in an act of what might be best described as depraved indifference for black life, pumped sixteen bullets into the adolescent's body, several as McDonald lay defenseless on the pavement.

Van Dyke was put on paid desk duty from October 21, 2014, until the day the video was released to the public. Just hours before its release, Anita Alvarez, Cook County prosecutor, finally charged Van Dyke with first-degree murder. Clearly, her decision was influenced by mounting public pressure: she had viewed the video more than a year earlier without taking action against Van Dyke.[11]

Predictably, Van Dyke's defense drew heavily on the "PCP-crazed black man" myth. One expert witness for the defense, James Thomas O'Donnell, claimed that PCP could cause "violent rage behavior" and could make a person feel as if he has "superhuman powers." Van Dyke told the court that McDonald's "face had no expression. His eyes were just bugging out of his head." As a result, he shot McDonald because he feared for his life. The jury didn't buy it. They convicted

Van Dyke of second-degree murder and sixteen counts of aggravated battery with a firearm, one for each bullet he fired.

According to Illinois statutes, Van Dyke faced a prison sentence of four to twenty years for the murder conviction alone. In addition, each aggravated battery count carried a minimum of six years behind bars. In short, Van Dyke should have received a prison sentence of at least one hundred years. But judge Vincent Gaughan discounted each of the aggravated-battery convictions and sentenced him to only six years and nine months in prison. Van Dyke is eligible for parole after serving about three years behind bars. This doesn't seem like justice. It's demoralizing and disgraceful.

It's hard for me to understand that a person, any person, could shoot another nonthreatening human sixteen times in cold blood. The fact that Chicago officials shielded Van Dyke from the consequences of his crime for more than a year still fills my heart with anguish. There is no way those in charge could have viewed McDonald as human, as deserving of humane treatment, as were they themselves and their loved ones. I think Jamie Kalven hit it on the head when he wrote, "Laquan McDonald—a citizen of Chicago so marginalized he was all but invisible until the moment of his death."

There were so many other people who worked tirelessly to ensure that Laquan McDonald was finally seen. For one, William Calloway filed the Freedom of Information Act request that forced the city to produce the dashcam video for public viewing. He made it possible for all of us not only to see Laquan McDonald but also to see the chilling disregard for a black boy's life displayed by some public authorities.

IN RECENT YEARS, psychedelic advocates have successfully pushed for wider mainstream acceptance of specific substances, while quietly dissociating themselves from others. For example, in 2019, Denver and Oakland passed measures that effectively decriminalized personal use of psilocybin mushrooms and other psychoactive plants and fungi. MDMA is expected to receive FDA approval for treating post-traumatic stress disorder (PTSD) within the next couple of years.

Meanwhile, psychedelic enthusiasts were conspicuously silent when Van Dyke used PCP as justification for his savagery. We also didn't hear a peep from them when Betty Jo Shelby, a white Oklahoman police officer, evoked the "crazy nigger on PCP" defense to justify her killing of unarmed black Terence Crutcher. On September 16, 2016, the day Crutcher was killed, he had PCP in his system, but a video of the incident clearly indicates that he was neither aggressive nor violent. Nevertheless, Shelby was acquitted of manslaughter charges. The legend of the superhuman, violent PCP user will live on. This means more people will die needlessly.

I am deeply disturbed that there is a deafening silence from the psychedelic community while fellow drug users continue to be brutalized as a result of PCP-related misapprehensions. The question before me is, Why the silence in the face of such egregiously harmful mischaracterizations of PCP? It might have something to do with the fact that black men bear the brunt of this murder-justifying myth. Or it could be that psychonauts are simply strategically protecting their mission to ensure continued public support for a select few psychedelics, including DMT, MDMA, and psilocybin. Drawing attention to the fact that PCP is also a psychedelic might jeopardize the reputation, and thus the availability, of other psychedelics.

It's also possible, however, that most people, including psychedelic advocates, don't know PCP is a psychedelic. Plus, the "crazed black on PCP" legend is so ubiquitous in drug education and popular depictions that its validity isn't questioned. Honestly, I don't know the exact reason for the lack of public engagement by the community when it comes to correcting misinformation about PCP. I hope the information presented here helps to change this situation.

I vividly recall lamenting this in a recent conversation with my friend Rick Doblin. Rick is the founder and executive director of the Multidisciplinary Association for Psychedelic Studies (MAPS). Over the past thirty-plus years, he has worked tenaciously to get psychedelic research studies approved and funded. He has teamed up with researchers—some of whom even consider him their mortal enemy—as well as therapists, patients, and activists in this pursuit. Few people on the planet know the FDA's drug-approval process better than Rick, who has a PhD in public policy and decades of experience working with and around regulatory agencies. MAPS, under Rick's leadership, has been the single most important driving force behind the recent broader acceptance of psychedelic, especially of MDMA, use. But the thing that stands out most about Rick is his ever-present beaming smile. If ever there was an eternal optimist, it's Rick.

In response to my frustration regarding the community's quiescence surrounding misguided notions about PCP and violence, Rick expressed similar concerns. But he also asked me what I had done to change or improve the situation. I had avoided attending conferences and events focused exclusively on psychedelics, for many of the reasons mentioned above, including the oppressive lack of racial diversity and pervasive drug elitism in those spaces. Still, Rick's question compelled me to acknowledge that I, too, was a member of the psychedelic community, and, as such, I had specific responsibilities, including providing education about issues that concerned me and should concern us all.

Rick challenged me to give a series of lectures at MAPS-sponsored events. Before I answered, he told me the story of how increasing the availability of psychedelics for adults became his life's mission. In 1972, Rick, who is Jewish, was an anxiety-ridden college student. While he was deeply worried about the possibility of being sent to war in Vietnam, he was even more terrorized by the atrocities carried out during the Holocaust. "The fact that people could dehumanize others and kill them," he told me. "That could be me one day." Rick went on to say that psychedelics helped him to see how we're all connected, to see the goodness in all humans. If others could also experience these types of insight, then perhaps, he believes, we would behave and live more compassionately. This belief is what motivates him every day to continue his mission.

I almost always come away from talks with Rick wanting to be a better person. Perhaps I had been unfair in castigating an entire community, I thought. I was reminded of the Maze song "We Are One." I thought about the song's lyrics that implores better treatment of each other and celebrates the joy that comes from it. I was determined to be better, to be more forgiving. True psychonauts, such as Rick, possess the values I want to emulate. So I agreed to do my part and to speak at his events whenever possible.

Rick's openness and nonjudgmental approach strangely reminded me of how I once smugly dismissed Jerry Garcia as someone unworthy of serious consideration in discussions about drugs, liberty, or happiness. I was so wrong. Garcia, contrary to my uninformed view, used not only psychedelics but also other drugs such as cocaine and heroin. Unlike some in his circle, loved ones and bandmembers, he neither looked down on users of nonpsychedelic drugs nor disparaged others for engaging in behaviors outside societal conventions, so long as these behaviors did not infringe on other people's freedoms. I wish more people in today's psychedelic movement would emulate these specific aspects of the life that Garcia tried to live. If so, they might actually gain an appropriate appreciation for the country's founding document—the Declaration of Independence— and for what Garcia meant when he said, "The pursuit of happiness. That's the basic, ultimate freedom."

9

Cocaine: Everybody Loves the Sunshine

Happiness lies within one's self,
and the way to dig it out is cocaine

Aleister Crowley

"Welcome to Colombia!" Those were the jubilant words that greeted me as I entered the VIP room of a Bogota nightclub. Neatly divided lines of cocaine were laid out on the coffee table in front of the woman who had delivered the greeting. I had come to the Columbian capital to give a lecture at the 2018 Psychoactive Week forum, which had finished only a few hours earlier. This was the afterparty. It looked like a scene straight out of *New Jack City* or some other awful cautionary tale intended to warn young people against cocaine use and trafficking. But things aren't always as they appear.

THE BRAZILIAN PARADOX

Nowhere is this more apparent than in Rio de Janeiro.

In 2013, I met Julita Lemgruber, a sociologist from Rio, at the International Drug Policy Reform conference in Denver. She urged me to come to her country and give a series of lectures. She felt that my views on drugs and society, especially my belief that specific substances are scapegoated in order to avoid addressing complex social problems, would resonate with Brazilians.

I was flattered but had no idea if this was actually true or not. I knew almost nothing about Brazil and even less about its approach to drugs. I was also self-conscious about my inability to speak Portuguese. I had spent years remedying other early educational deficiencies, but the acquisition of other languages remained a glaring gap in my skill set. I didn't want to subject myself to feeling like "the dumbass American" who couldn't be bothered to learn a foreign language. I politely declined Julita's invitation.

But Julita isn't the sort of person who takes no for an answer. She is persistent, persuasive, and tough. Undoubtedly, these attributes contributed to her being selected as the first woman in charge of the state of Rio's prison system back in the early 1990s and to being named the state's first police ombudsman from 1999 to 2000. Julita doesn't look the part in the traditional sense. She sports fashionable gear and rocks a stylish crimson pixie cut with straight bangs. But behind her youthful face lies decades of experience, disappointment, and wisdom.

Julita is her own person; she's a fearless, independent thinker who's unafraid to go where the evidence dictates. Julita's perspective has continually evolved over the course of her long career working closely with law enforcement. She's now convinced that draconian drug policies and racial discrimination are at the center of the crime and violence she has spent much of her life trying to stem. This belief fuels her unwavering efforts to push for less restrictive drug laws and for the meaningful inclusion of marginalized people into Brazilian society. The more I learned about Julita, the more difficult it became to say no.

So, in May 2014, there I was, in Rio, lounging at a luxury hotel that overlooked the famous Ipanema Beach. When Saturday rolled around, I was asked to take part in the local annual Marijuana March. But the march's exclusive focus on cannabis, as if it and its users were higher on the pecking order of drugs, gave me pause. In the end though, I participated and walked the entire beach from one end to the other. I walked alongside Jean Wyllys,[1] a local celebrity and politician who had introduced bills to legalize marijuana. Hundreds of others joined us. It was a community of kindness brought together by a desire to legalize adult recreational marijuana use. The vibe reminded me of Roy Ayers's 1976 hit "Everybody Loves the Sunshine." It was cool; it was festive, too. People openly sold, shared, and smoked weed. They also exchanged other things—food, drink, love, you name it, and it was permitted, or so it seemed.

Over the course of the day, I was told that *all* drugs had been decriminalized in Brazil since 2006. This surprised me. I assumed that Brazil blindly followed the United States when it came to drug policy. Wrong. According to Brazilian law, unlike U.S. law, anyone caught possessing substances in amounts consistent with personal use should not be subjected to incarceration. Instead, the person might receive a warning and be required to perform community service or attend a drug-education program or course. Those caught selling banned drugs, however, were still subjected to harsh criminal sanctions.

The morning after the Marijuana March, I met with Julita in a restaurant in Leblon, an upscale neighborhood near Ipanema Beach, to discuss the schedule of events she had prepared for me. It looked exhausting, daunting even. I was to travel across three states while giving multiple lectures and media interviews, making a number of site visits, and holding meetings with interested parties in each state—all in the course of about a week. I got tired simply looking at the itinerary. This was typical Julita—detail oriented, even if it kills her, and efficient, even if it kills you.

I was too embarrassed to express any apprehension I felt about the formidable task she'd laid out for me. I just had to grin and bear it.

I did, however, say something about what I had learned on the previous day at the march. I heaped praise on Brazilian lawmakers for passing such a progressive drug law. "My Deearrrr," Julita said very slowly in a tender tone. "Oh shit," I thought. I realized my mistake even before she opened her mouth. Praising politicians is a tricky endeavor because most will eventually disappoint you, especially when it comes to drug policy. But it was too late. The deed was done. Julita was now in the middle of telling me something of great importance, and there was no stopping her. Locking her kind but intense eyes firmly on mine, she recited Brazilian composer Antônio Carlos Jobim's famous quip: "Brazil isn't for beginners."

My education on Brazil began at that moment. Julita spent the next thirty minutes schooling me, practically nonstop, on how drug policy *really* plays out in her country. It was certainly true that under the current law, *personal drug use* is not meant to be punished by incarceration, but it remains a criminal offense. So, in effect, the law is *depenalization*, not *decriminalization*. Equally important, the law does not quantify personal use. It does not define drug amounts in terms of how much is considered personal use versus how much is considered trafficking. This critical factor is determined first by street-level police officers, who decide who is arrested and who is not. Arrestees, regardless of the drug quantity they possess, are eventually tried in criminal court as drug sellers.

Sure, the judge can ultimately rule that the defendant is not a trafficker, after taking into account the amount of drug possessed, the person's legal history, and other mitigating factors. But that ruling is almost never made, especially if the defendant is black and poor. Another less frequently discussed feature of the Brazilian drug law is that it increased the minimum amount of prison time for trafficking violations from three to five years.

To cut to the chase, Brazil's supposedly progressive drug law has actually dramatically *increased* the number of individuals in jail for drug trafficking. For example, drug arrests now account for nearly one-third of all arrests, whereas in 2006, when the law was passed, drug arrests accounted for about 10 percent of arrests. Furthermore, evidence shows that most individuals convicted of drug trafficking are unarmed first-time offenders with small amounts of drugs. And judging from prison population demographics, African-Brazilians are bearing the brunt of these arrests. While they make up about half the general population, they account for 75 percent of prison inmates.[7] The message seems to be that if you're white, you're a user. You can go home. But if you're black, you're a trafficker. You must go to jail. And you can remain there for several months without ever appearing before a judge.

"The law is only as good—or as fair—as those interpreting it," Julita said, the frustration palpable in her voice. She told me to take a look around the restaurant. "How many black people do you see?" she asked. I was the only one. Toni Morrison's extraordinary novel

Paradise came to mind. In it, the author noted that paradise is "defined by who's not there, by the people who are not allowed in."[3]

In August 2015, I had an experience that made this point personal. By then, I had visited Brazil several times, and I had learned
quite a bit about the ongoing discrimination against the poor, especially those defined as black Brazilians. It's important for me to
stress that my status as an Ivy League professor and scientist provided some protection against antiblack Brazilian racism. During
this August visit, it was reported that I was denied access to one of
São Paulo's five-star hotels, where I was scheduled to deliver a lecture.[4] Within twenty-four hours, the story went viral, inciting a public outcry of rage and embarrassment that a black man was denied
entrance to a hotel because of his race. I received an outpouring of
support. Hundreds of people sent their encouragement and apolo-
gies via social media and email. I couldn't walk the streets of São
Paulo without someone stopping me to express their sympathy and
sadness for how I was treated by the hotel staff.

Fortunately, the story wasn't true. I was never denied entry by
the hotel staff. I was completely unaware of the extent of this drama
until I read about it online. What I found particularly disturbing is
this: the egregious racial discrimination that occurs daily in Brazilian society does not generate a fraction of the attention, sympathy,
and guilt that this bogus event did.

Consider two blatant examples of racism that occurred that very
same week. In one case, it was revealed in the press and confirmed
by the government that the police had been removing groups of
black boys from public buses in an effort to prevent these children
from going to Ipanema and its surrounding beaches. None of the
children were charged with a crime; nonetheless, the policy was justified as a crime-prevention strategy. A remarkably large segment of

Rio's residents support such racially discriminatory measures. I have yet to hear that any public officials apologized to these black boys for this mean-spirited and shameful policy.

Another case involved a protest in response to the August 14, 2015, massacre of nineteen people, almost all black, by a clandestine police squad said to have been avenging the deaths of two fellow officers. Unbeknownst to me, the protest took place a few blocks from my hotel on Friday, August 28, the same day I gave a lecture to a group of criminal lawyers. Sadly, at least four times as many people attended my lecture as took part in the protest.

Initially, I was puzzled by the tremendous amount of public interest in the alleged racial discrimination perpetuated against me. It's now clear, however, that the press and public are far more comfortable focusing on individual acts of racial discrimination—especially when the victim is an American public figure—than on ongoing racial discrimination against voiceless ordinary citizens. There are countless examples of this phenomenon, in both Brazil and the United States.

The more time I spent in Brazil, the more parallels I saw with my own country. Some emerged with cruel clarity. In both countries, for example, the opinion makers and authorities—including politicians, journalists, law-enforcement personnel, and educators—are exquisitely adept at using drug-related issues to denigrate and subjugate black and poor people. Exaggerations of issues surrounding crack cocaine epitomize this phenomenon. The drug has been blamed for both countries' most vexing social problems, ranging from high rates of black unemployment to inhumanity to criminality.

Politicians get away with simply scapegoating the drug *du jour* in order to avoid addressing the real problems people face: poor education, an insufficient number of jobs paying a living wage, affordable housing, racial discrimination, and a lack of basic public services, to name just a few. This isn't news, however. Politicians know that it's far more politically expedient to offer what looks like immediate solutions to trumped-up drug crises, such as hiring more cops, than it is to invest in appropriate social policies whose benefits may not be seen for several years after the election cycle.

CRACK IN THE UNITED STATES: THE EPIDEMIC

Around 1985, crack became widely available in major U.S. cities. By the late 1980s, it was being blamed for everything from black unemployment to misrepresented high murder rates to crack babies. The problem with that line of argument is that per capita murder and unemployment rates were higher in 1980 and 1982, respectively, before the introduction of crack. We also now know that the whole crack-baby craze was wildly overstated.[5]

But why let facts derail a good story? The crack tale was going to be played out just as other drug scares before it had, no matter what the evidence said. By frightening the population about the dangers of a purportedly new drug, cultural moralists are provided with enticing opportunities to impose their views on society. The moralists "help" to delineate clear lines between good and evil—never mind that we know that people are neither entirely good nor entirely evil. The moralists promote an *us*-versus-*them* mentality—never mind that we know this causes dangerous tensions between groups.

I grew up in Miami in an all-black area with few economic opportunities. Outsiders characterized our neighborhood as lawless and particularly unsafe for nonblack people. As I mentioned earlier, the prevailing sentiment in the late 1980s was that crack dealers and addiction were the causes of all that ailed my hood. The same was said about many other black communities. Conventional wisdom held that the drug was so virulently addictive that users needed only one hit and they were hooked for life. In a widely read exposé about crack dealers, Barry Michael Cooper wrote in the progressive *Village Voice* that because of crack, "outlaw is the law."[6]

Even I got duped. In fact, I decided to study neuroscience specifically because I wanted to cure crack addiction. I also low-key joined Nancy Reagan—and other cool-ass celebrities, such as Pee-wee Herman, who connected with young people—by promoting slogans that urged folks to "Just Say No" to drugs. And who wasn't a fan of Keith Haring's huge "Crack Is Wack" mural located in New York City off Second Avenue and 128th Street?

Frankly, I think the crack issue gave pedantic, pseudointellectual, conscious brothas, such as twenty-one-year-old me, a *raison d'être*—not to mention career prospects. I received mad praise, attaboy, for advising black youth to stay off drugs. It was affirming. It felt good, like I was doing something important. And although I didn't possess the skills to verbalize this back then, I knew that potential white employers are far more comfortable hiring black men to police the behavior of other black men than hiring them to serve in other capacities. It's not an accident that a substantial proportion of low-level security positions are occupied by brothas.

For politicians, crack was a wet dream. They used the issue to justify waging an even more intense war on drugs. Congress passed new laws claiming to protect "real" Americans from unsavory drugs, dealers, and doers. As I've mentioned, one law, the Anti–Drug Abuse Act of 1986, established penalties that were one hundred times harsher for crack infractions than for powder infractions. Another law, the Anti–Drug Abuse Act of 1988, even promised a "drug-free

America" by 1995. This goal—spoiler alert—wasn't achieved. But this didn't deter Congress from markedly increasing the budgets of law enforcement year after year in the antidrugs effort. Predictably, the crack frenzy led to record numbers of people jailed for drug-law violations. It set in motion the era of mass incarceration. More than two million Americans will sleep behind bars tonight.

The discourse on crack was unabashedly racialized. The enforcement of federal crack laws is but one example of this deplorable practice. A whopping 85 percent of those sentenced for crack offenses were black, even though most users and dealers of the drug were, and are, white. Without a doubt, this type of racism has contributed to spine-chilling statistics, such as this one: despite making up only 6 percent of the general population, black males comprise nearly 40 percent of U.S. prisoners.

The race-and-pathology narrative in which crack was steeped permeated the media and popular culture, too. The 1991 film *New Jack City* is emblematic of this phenomenon. The film chronicles the rise and fall of Nino Brown, a fictitious black crack druglord portrayed by Wesley Snipes. According to the story, Nino was bright enough, charismatic enough, industrious enough, persuasive enough, and ruthless enough to take control of an entire NYC public-housing project so he could set up the most profitable crack business the city had ever seen. Shit, at the time, even I secretly wanted to be Nino.

New Jack got rave reviews. According to critic Roger Ebert, the film provided Americans with "a painful but true portrait of the impact of drugs [crack and powder cocaine] on this segment of the black community."[7] Right on, brotha Roger! You nailed it. *New Jack* confirmed my views about crack. Probably because, basically, the filmmakers just dramatized the sensational media reports I had read about the drug. So, naturally, Ebert's words resonated with me: "We see how they're [drugs are] sold, how they're used, how they destroy, what they do to people."

CRACK IN THE UNITED STATES: THE EPIDEMIC THAT WASN'T

Years later, I realized we'd all been horribly wrong. We'd not just been wrong to have adopted those views on crack; we'd also been wrong—deplorably so—to have heedlessly dehumanized those who sold or used the drug. This allowed authorities to shift the focus away from a "war on drugs" to a "war on people." To put it more bluntly, it was a war on my people. I still haven't gotten over the profound regret I feel whenever I think about the ignorant, traitorous role I played in vilifying crack and the people who were targeted.

I hope my current work as an academic and scientist helps to set the record straight. I have given thousands of doses of crack to people as part of my research and have carefully studied their immediate and delayed responses without incident. Sure, the drug can, in rare cases, exacerbate preexisting cardiovascular problems. But, in general, its cardiovascular effects are comparable to those that occur when people regularly engage in intense exercise.

Contrary to popular belief, the effects produced by crack are predominantly positive. My research participants consistently report feelings of well-being and pleasure after taking the drug. Pleasure is a good thing, something that should be embraced. It feels weird that I am compelled to write the preceding sentence because the idea seems so obvious. But I know that there remain those who obstinately cling to the belief that crack-induced pleasure is so overwhelming that it drives most users to uncontrollable consumption.

The data say otherwise. The addictive potential of crack—or the latest vilified drug—is not extraordinary. The fact is that nearly 80 percent of all illegal-drug users use drugs without problems such as addiction.[8] In other words, we now know unequivocally that the effects of crack have been ridiculously exaggerated; crack is no more harmful than powder cocaine is. They are, in fact, the same drug.

I have published these and related findings in respected scientific journals as well as in popular outlets.[9] I have given countless public lectures and media interviews dispelling myths about crack and drawing attention to this particularly somber fact: the egregious popular portrayals of crack have ruined more lives than the drug itself.

TOM THE LUMBERJACK FROM IDAHO

Our response to crack in the 1980s and 1990s weighed on me even more heavily in the summer of 2015 as I worked in a Geneva clinic, where people diagnosed with heroin addiction received multiple daily doses of the drug as part of their treatment. Sometimes, while I watched the patients obtain and inject their medicine in this comfortable and respectful environment, I couldn't help but think about the contrast between the Swiss's approach to heroin and our own to crack, or any banned drug for that matter.

One day while in the clinic, I received an email from someone named Tom Wright. He claimed to be the *New Jack City* writer. Cheekily, he asked something like, "Why are you dissing my movie?" Sometimes when I spoke publicly about cocaine, I'd mock the movie as unrealistic and harmful. But I didn't think the filmmakers would have the nerve to step to me. "This must be a prank," I thought. Nope, it wasn't a joke. It was, indeed, *the* Tom Wright, Mr. *New Jack* himself.

After a few friendly emails, we set up a face-to-face meeting upon my return to the United States. We'd connect at the Columbia campus entrance on Broadway and 116th Street and then walk to a restaurant for lunch. As I strolled toward our meeting place, an anxiety-provoking thought ran across my mind: "Damn, how will I recognize Tom?" I'd never met him, and the Broadway entrance is always overrun with visitors. "Ah yeah," a second sinking thought pushed out the first, "whites and Asians are the overwhelming majority of people entering the campus." This meant that the *New Jack* writer would likely be easy to spot.

He wasn't. Despite the fact that he stood only a few feet away, several minutes passed before I recognized him—in part, because I was trying to avoid getting waylaid by the flannel shirt–wearing, middle-aged white guy who had just waved at me. Perhaps he recognized me from TV or something. I couldn't be sure. But I wasn't going to make eye contact with him because it might've been perceived as an invitation to engage. Ignoring my uninviting demeanor, flannel-shirt guy introduced himself. It was Tom. I was expecting a black man, someone who dressed like Teddy Riley or the Cash Money Brothers. I couldn't have been more wrong. Tom resembled a lumberjack far more than he did a new jack.

Later, we joked about my mistaken perception of him. He said it happens often, in part, because that's the way *New Jack City*'s producers wanted it. When the movie premiered at the Sundance Film Festival back in January 1991, Tom was asked to stay home. Apparently, it's difficult to market an "authentic" film about black urbanites and crack if the screenwriter is white and from Idaho.

The folks at Warner Bros. were well aware of this image problem from the moment they acquired Tom's story. To address it, they hired Barry Michael Cooper to massage the story line. Cooper, a black writer from New York City, certainly had more "street cred" than Tom did. In addition, Cooper had already written an influential article on crack titled "Kids Killing Kids: New Jack City Eats Its Young."

Published in *The Village Voice* in December 1987, the lengthy essay detailed Detroit's—but, oddly, not New York City's—so-called crack epidemic. In the piece, Cooper introduced mainstream America to the term *new jack*: "a calculated novice who enjoys killing you, aside from making a name for himself." Those he labeled as new jacks were not just the people in the crack game—which is bad enough—but also black youth in general. He lambasted my contemporaries for everything from their desire to have disposable income to the gear they wore to the rap music they listened to. Drawing heavily on sensationalism, he told a story that blamed young black Detroiters for much of the city's chaos. As you might have guessed, given the times, Cooper called for tougher drug laws. They were especially needed, he claimed, because my peer group was uniquely "immune to the harsh punishment for drug trafficking."

Americans from all walks of life fell hook, line, and sinker for this dehumanizing depiction of black youth. The producers of *New Jack City* couldn't lose with Cooper on board as a storyteller. He reworked Tom's original screenplay to more closely match his *Village Voice* article. He changed the drug of focus from heroin to crack. The screenplay title was also changed: *The Godfather: Part III* became *New Jack City*.

Why was the original version called *The Godfather: Part III*?

Tom was initially hired by Paramount Pictures to write the final screenplay of *The Godfather* trilogy. But there was a problem: the star of Tom's flick would be a black man. In the 1970s, the film's backdrop, Nicky Barnes was one of the most prominent figures of the New York City underworld. So naturally he'd be featured in Tom's screenplay, right? Wrong. There was no way the higher-ups at Paramount were going to make *The Godfather* film with a black person in a high-profile role. To make matters worse, Eddie Murphy, then Paramount's box-office cash cow, really wanted to play the Barnes character. The top brass, however, feared that Murphy's movie-star image would be tarnished, possibly irrevocably, if he portrayed a drug dealer. So rather than developing Tom's original screenplay, Paramount allowed Tom to shop it to other studios, thereby killing two birds with one stone. Quincy Jones, then at Warner Bros., scooped it up, assembled a team, and the rest, as they say, is black history.

During our conversation, I got the feeling that Tom had had practically no editorial influence on what would become *New Jack City*. "Boy, are you lucky!" I thought as he continued. He had been shut out of the film's promotional activities, which meant he was virtually anonymous. Few people associate his face with the film. That's exactly how I'd like it if I were him. I certainly wouldn't tout my role in creating anything that relies on extreme distortions to reinforce misguided notions that black youth are uniquely prone to descend into savagery.

Today, in the United States, crack is no longer considered the worst drug in the history of humankind. That distinction now—for the time being—belongs to opioids, but in a few years, you can bet it'll belong to another drug. As for crack, many acknowledge that exaggerations about the drug drove us to adopt preposterous poli-cies, which, among other things, contributed to the further margin-alization of black people. In fact, on August 3, 2010, President Obama signed the Fair Sentencing Act, which reduced the sentencing dis-parity between crack and powder cocaine from 100:1 to 18:1. This was an important acknowledgment of our past foolishness, but, to be clear, any sentencing disparity in this case makes no scientific sense.

NEW JACK RIO

And yet, despite the dreadful and lingering impact of U.S. crack policies, Brazil is pursuing a similar path some thirty years later. Many Brazilians have been convinced that "cracolândias" are one of their country's most pressing problems.[10] Supposedly, crack lands—the English translation—are places where "fiends" gather to smoke the drug, as well as engage in other behaviors that offend the domi-nant culture. (In the United States, we once called these places "crack houses.") Located in urban slum areas, cracolândias are also reputed to be controlled by young druglords who use deception, coercion, and violence to "hook" users in order to guarantee loyal long-term customers. Some claim that cracolândias are a major source of African-Brazilians' undoing. Sound familiar?

I know that most stories about crack don't square with reality. That's why one of the first places I visited in Brazil was a cracolândia. I was warned that these places are filled with unpredictable "zom-bies" driven primarily by their desire for another hit. "Barbaric" was the term used to describe them to me by one person who advised

against my going. I got similar reactions about my plans to visit favelas, districts long abandoned by the government but home to many
of the country's poorest citizens. In favelas, the state often does not
provide basic public services, such as medical care, sanitation, and
transportation. The void is usually filled by community members
themselves, evangelical churches, NGOs, and, yes, criminal organizations. Cracolândias and favelas share some important characteristics. Both are inhabited primarily by people on the margins of society
who suffer from incendiary and inhumane perceptions constructed
mostly by outsiders and who dwell in precarious conditions where
misery, illegality, and violence predominate. Cracolândias are typically located in favelas.

A bumper sticker on the back of a police car caught my eye en
route to my first visit to these areas. It read in Portuguese, Crack, É
Possível Vencer (Crack, It's Possible to Win). I dismissed it as just
rhetorical drug-war propaganda, not thinking it was promoting an
actual war. I was dead wrong. When we arrived at Complexo da
Maré, one of Rio's largest favelas, I saw an actual war zone. Armed
forces were everywhere, an occupying force to the favela's 140,000
residents.

The official story claimed that the military was needed to restore
order and halt the violence caused by the crack trade. Crack was the
enemy, and it would be defeated. Others say that this is a cover
story, that the troops were deployed at Maré because of the upcoming 2016 Rio Olympics. Maré is located on the main highway
leading to and from the city's major airport. Officials worried that
unsavory activities might've oozed out of the favela and into the
global spotlight. Rather than risk potential embarrassment, the decision was made to bring in armed forces.

When I arrived at Maré for the first time in May 2014, I was taken aback by the imposing number of troops patrolling the favela. I'd never seen anything like it, and I had spent time in the military. But apparently, the army soldiers were also surprised by our presence there. I was with a group of bourgeois-looking white Brazilians. And we stood out like Europeans on a Kenyan safari, complete with wide-eyed wonder and high-performance cameras. Slightly annoyed, but definitively confused, the soldiers demanded to know why were we there. They had the power to turn us away, to make us leave. They knew all too well that most middle-class Brazilians avoid favelas like the plague, which meant, among other things, that the soldiers had *carte blanche* to carry out inhumane acts with impunity and without the prying eyes of the press.

I vaguely remember someone in our group telling the soldiers that I had come from the United States to speak with kids from favelas about staying off drugs. One of the armed men radioed this information to a superior who eventually allowed us to continue on our way.

The whole encounter was surreal, chilling even. We stood only a couple of meters away from a dozen or so heavily armed soldiers, seemingly adolescents. Many of them grew up and still live in the very same favelas they occupy. I thought back to the time when I was in the military, when I was their age, carrying a loaded automatic rifle. Like them, I did as I was directed. Thankfully, I was never ordered to subjugate my own community. I felt for the soldiers; they were just kids. I felt for the residents; hundreds are killed each year by men in uniform, mostly police.

In Brazil, it's not always easy to tell the difference between the federal armed forces and the state police. Both routinely invade favelas. Regarding the state police, it's important to understand that they comprise two types of forces: civil and military police. The civil police are responsible for criminal investigations, that is, detective work, forensics, and prosecutions. The military police, on the other hand, are organized like the federal armed forces. In fact, each member concurrently serves as a reservist in the Brazilian army and receives training in counterinsurgency. Military-police units are also equipped with armored vehicles and high-powered automatic assault rifles. These units' sole charge is to maintain public order, which includes frequent occupation operations. The problem is that authorities sometimes deem routine life in favelas as "public disorder," justifying, in their minds, an invasion. In Brazil, the military police are frequently used as an invading army against the country's own poor citizens.

I went to several favelas and so-called crack lands. I indeed saw people smoking crack out of makeshift pipes and drinking alcohol out of plastic cups. I saw heated and animated discussions. But I mostly saw people talking, laughing, and tending lovingly to their children and pets. I saw people living life.

Mostly I saw the widespread abject poverty. A large number of people lived in shoddily constructed shacks, devoid of basic services and surrounded by piles of rubbish. It seemed that the authorities had not removed the trash in some of these communities for months. I grew up in a housing project and yet was still absolutely shocked and disturbed by these conditions. I tried not to show my dismay, because I was also grateful to be in the presence of such decent and generous people.

The residents were extremely warm and welcoming. So-called drug users and traffickers were eager to share with me. At my request, one person even gave me a crack rock to have tested for purity; sadly, I couldn't find a testing site in the country. Some told

stories of male relations being rounded up by the police for suspected drug trafficking, never to be seen alive again. Residents
didn't need to be told that problems such as widespread poverty,
inferior education, high unemployment, and violence had plagued
their communities long before 2005, when crack first appeared in
Brazil.

This is borne out by the data. Brazil has been beset with high,
usually double-digit, rates of unemployment since it became a democratic country in 1988. Unemployment peaked in the late 1990s at
just under 15 percent. Brazil's unemployment rate is usually more
than twice as high as that of the United States'. Homicide rates in
Brazil have been consistently among the highest in the world for
several decades. Between 1990 and 2003, the countrywide rate increased from 22 to 29 per one hundred thousand residents. This
increase was followed by a slight decline to 27 in 2011, only to peak
at 31 in 2017. In 2018, the number dropped to 25, still five times
higher than the murder rate in the United States.

The popular rhetoric is that drug gangs are largely responsible
for the social instability and violence in Brazilian urban centers,
such as Rio. Politicians often invoke this claim to justify the tanks
and soldiers that have become commonplace in some favelas. Dressed
in combat fatigues, the police are at war with the country's poor
and black citizens, a war fought in broad daylight, in a democratic
society.

Heavily armed local militias have become a normal feature of
favela life. Comprised mainly of off-duty and retired police officers,
militias came about supposedly to protect favela residents from drug
traffickers. In truth, they operate much like the criminal organizations they claim to keep in check. They clash with traffickers over
control of lucrative regions. They extort money from residents and

shopkeepers. And they sell drugs. In Rio, militias control almost half of the city's nearly one thousand favelas. By comparison, drug traffickers control less than 40 percent.[11]

In 2018, police killed more than 6,100 people nationwide. That's about six times the toll in the United States, whose population is larger than Brazil's by 115 million people. Many of Brazil's police killings amount to extrajudicial executions, just as in the Philippines. The São Paulo police ombudsman examined hundreds of police killings in 2017 and concluded that excessive force was used—sometimes against unarmed people—in three-fourths of the cases.

The largest proportion of police killings tends to occur in the state of Rio, which is also the home of the country's notoriously callous president Jair Bolsonaro. From 2003 to 2018, on average, Rio police killed 930 citizens each year, 70 percent of African descent. In 2018, this number soared to 1,534; more than four people were killed daily at the hands of the police.[12] Halfway through 2019, the average number of citizens killed by officers had climbed to more than five per day.

Many people, me included, see these police killings as a campaign of genocide. But not Bolsonaro and his supporters. Consistently, he makes public comments that demonstrate a blatant disregard for due process and that encourage police brutality. Suspects should be shot dead "in the streets like cockroaches," he has said. It is no wonder that Jean Wyllys, a Brazilian politician, feared for his life and fled the country shortly after Bolsonaro was elected president. Wyllys was a noted Bolsonaro nemesis when they both served in the Brazilian congress.

Bolsonaro's barbarism might be surpassed by that of Wilson Witzel, governor of Rio. A former judge, Witzel is known for urging the

police to "aim at their little heads and fire!" when dealing with suspects.[13] With these types of leaders, an end to the country's long
history of social instability and violence against specific groups does
not appear to be anywhere in sight.

Despite the complex mix of factors that contribute to the country's pressing problems, too many Brazilians aim to address them
by starting first with crack users and drug traffickers. Like U.S. officials more than thirty years ago, Brazilian authorities feel justified
slaughtering poor brown and black people so long as it's in pursuit
of "public security." This means, among other things, eradicating
crack users and dealers, whatever the collateral damage. The familiar script—frighten the public about the violent unpredictability of
crack users and traffickers—allows authorities to divert attention
away from legitimate concerns and increase the budgets of law enforcement and "treatment" providers.

In 2014, the country allocated R$4 billion in this effort. Public
awareness and education campaigns are included, although what
parades as education cannot be considered informative. Drug education amounts to telling people not to take illegal drugs. In Brazil,
drug treatment primarily consists in mandating users to facilities
run by evangelical Christian organizations, where the focus is on
prayer and manual labor. By any modern standard of medicine,
this can hardly be considered treatment, let alone effective treatment. The bulk of the funds and focus of Brazil's crack efforts is
geared toward law enforcement, just as is the case in the United
States.

This will undoubtedly lead to more African-Brazilian deaths and
push these folks further to the margins of society. African-Brazilians
make up about 50 percent of the population but represent less than

5 percent of elected officials and are virtually nonexistent in middle-class positions.

So what should be done in Brazil? That's a complicated question, with answers way beyond the scope of this book. But meaningful efforts to increase educational and economic equity would go a long way. Another solution would be to dispense with the scapegoating of crack, or any form of cocaine. Anyone who believes that crack—or any drug, for that matter—is the major problem faced by marginalized people is either dishonest or naive, or both. If anything, cocaine provides some respite from the suffering among the poor and from the cognitive dissonance experienced by conscientious, well-to-do white Brazilians, who know that what is happening in their country is obscene.

SUNSHINE TO BRIGHTEN MY DAY

Back in Bogota, I sat at the cocaine-filled table, half listening to the chemist who was demonstrating how to distinguish between high- and low-purity cocaine. Each line contained varying percentages of the drug, ranging from about 20 to 90. He knew this because earlier he'd analyzed the samples in preparation for the demonstration. I remember this gentle hombre referring to cocaine as "sunshine to brighten up your day." I also vaguely recall him saying that you can get a pretty good estimate of purity by checking the wetness of the substance. The wetter the cocaine, the better the quality. At least, that's what I think he said.

Honestly, I found it difficult to pay attention. My mind had drifted to Brazil and its hypocrisy about cocaine. I fixated on the

"Helicoca Incident," when on November 24, 2013, the Brazilian federal police seized a helicopter filled with a half a ton of cocaine. The helicopter belonged to a company owned by the family of Brazilian senator Zezé Perrella. At the time, his son Gustavo was a representative for the state of Minas Gerais. Gustavo used part of his government allowances to fuel the helicopter and employed the pilot as a personal assistant. Despite these connections, neither of the Perrellas were prosecuted. The pilot took the hit and was convicted of drug trafficking. He got a ten-year bid; the Perrellas got their helicopter back.

A similar incident occurred in June 2019, when a member of President Bolsonaro's military detail was caught with 39 kg of cocaine on the way to the G20 submit in Japan. During a stopover in Seville, Spanish authorities found the drug in the handbag of Silva Rodrigues, an airman in the Brazilian air force. Bolsonaro was traveling on a separate plane that did not land in Seville. He said in a statement, "If the airman is found to have committed a crime, he will be tried and convicted according to the law." Rodrigues was the only person arrested and remains behind bars in Spain. It looks as if he's the fall guy in this case.

Thinking about the large numbers of poor and black Brazilians jailed and killed each year in anticocaine efforts overseen by depraved individuals—who, by the way, enjoy the drug as much as the next guy—made me feel hopeless, even complicit. How was I any different from liberal white Brazilians who secretly use their cocaine—and other drugs—without publicly aligning themselves with condemned drug users?

"How about a li'l sunshine?" the chemist asked, observing that my mind was elsewhere. His question drew me back into the room and provoked one of my own: "What kind of guest rejects such hospitality?" Certainly not me. My mother had raised me better, raised me to have manners. After the cocaine gently caressed my nose and the effects were apparent, I heard Bill Withers singing in my head: "Ain't no sunshine when she's gone." Perhaps we'd love each other better if we had more sunshine in our lives. We have a long way to go.

10

Dope Science:
The Truth about Opioids

Love is dope . . .

Tom Robbins

When I started writing this book, the verdict on opioids seemed clear cut. These drugs ravaged large portions of the country, causing immediate addiction and vast numbers of unexpected overdose deaths. I soon learned it wasn't that simple. As was the case with crack cocaine decades earlier, the story of opioids is far more complex than we have been led to believe. I ask that you read this chapter with an open mind and allow the evidence to determine your perspective.

• • •

"BLAHHHH!" THE RETCHING sound caused Robin to rush into the bathroom, where I was kneeling in front of the toilet as if it were an altar. "Are you OK?" she asked, a worried look on her face. Three weeks earlier, I had told her I was going to conduct an experiment on myself, during which I'd deliberately go through opioid withdrawal to see what it was like. This was it, fall 2017.

Only a few years earlier, I couldn't have imagined taking opioids on a consistent basis, let alone voluntarily undergoing withdrawal from them. I was too afraid. Media reports suggested that a person could get hooked after only a few hits. And once addicted, that individual ran the inevitable risk of dying either from an overdose or from opioid withdrawal symptoms. Who needed that? Certainly not me.

In 2014, I gave a book talk in Geneva, Switzerland, that focused on methamphetamine use. During the question-and-answer period, I managed to say a few uninformed things about heroin, despite their being outside the scope of my talk. I said something like "Chronic use of the drug undoubtedly produces physical deterioration; it damages your body." The fact that I had no evidence for this assertion didn't inhibit me. The statement just felt like it should be true, and it was in line with my own biases about heroin.

Immediately after the talk, I met Barbara Broers, who was in attendance. Barbara is a professor at the University of Geneva and an internist whose specialty is treating drug addiction. For several years, she worked at a Geneva clinic where heroin addicts are given daily doses of the drug as part of their treatment, in much the same way people take daily doses of a beta-blocker or insulin to control symptoms related to hypertension or diabetes.

Barbara explained that she wanted to learn more about my perspective on the issues I raised in my talk. She invited me for a hike on Mont Salève the following morning. I accepted, even though I didn't have the proper clothing or shoes. It was winter and cold, and snow was everywhere. To make matters worse, Mont Salève's peak stood nearly a mile above sea level. Having only recently arrived from New York City, I knew I'd be winded quickly if I exerted myself too much. But my ego overrode my reason. How bad can it be? I thought. If Barbara can handle it, so can I.

I quickly realized that it was I who would be doing the learning, not Barbara. It was clear that I was out of my league on multiple levels. Barbara is a serious athlete, although her modesty prevents her from saying so. She doesn't own a car but walks, jogs, and cycles everywhere. As we hiked, her stamina was evident: she talked a mile a minute without any signs of fatigue or shortness of breath. She also patiently and carefully listened. I, on the other hand, gasped for air like a fish out of water as I tried to keep pace both physically and intellectually.

"Heroin is one of the safest drugs," she said in a quiet, matter-of-fact tone. She qualified her statement with the usual pharmacological considerations, such as the need to carefully attend to the dose administered and the user's level of tolerance. I'm not exactly sure what I said or if I even said anything, but I am certain that the incredulous look on my face communicated, "Get the fuck out of here!" Barbara began by telling me about her experience working with patients in the heroin clinic and how well these folks do with its treatment. Many of them are also afflicted with other illnesses, including psychiatric disorders. She stated that heroin, compared with antidepressant and antipsychotic medications, has far fewer side effects. OK, I could see that; many psychiatric medications have numerous serious side effects. In some cases, the adverse effects are so debilitating that patients refuse to take those medications.

I also knew that heroin is made by adding two acetyl groups to morphine. This minor modification of morphine allowed Bayer Laboratories—yes, the same people who gave us Bayer aspirin—to market heroin as a nonaddictive cough suppressant. It was 1898, and there were minor concerns about the risk of physical dependence caused by the then more commonly used antitussive drugs, morphine and codeine. Both of these drugs are derived from the opium poppy and, like heroin, belong to the class of drugs called opioids. We know now that all opioids, including methadone, oxycodone, and fentanyl, are capable of producing physical dependence. At the time, though, this effect had not yet been observed with heroin, so it seemed an ideal replacement. Eventually, this view would evolve and the medical uses of the drug would be restricted primarily to pain relief. Today, heroin is used medically in several countries, including Ireland and the United Kingdom, but not in the United States.

Barbara continued schooling me. She said one of her most consistent clinical observations was that heroin is more effective at controlling psychotic symptoms, such as hallucinations, in many patients than are traditional antipsychotic medications. Initially, this was a bit much for me to digest, but once I got over my shock, I could at least see how this might be theoretically possible.

Antipsychotics are the drugs used to treat schizophrenia and other psychotic disorders. The dominant theory purports that psychotic behaviors, including hallucinations and delusions, are caused by overactivation of dopamine cells in the midbrain. Antipsychotic drugs block dopamine receptors and thereby prevent excessive dopamine activity. Supposedly, these drugs eliminate voices in the heads of schizophrenic patients and reduce their delusions. In reality, it's not that simple. Antipsychotics are not a cure. Many patients report that these drugs don't necessarily rid them of the voices in their heads, but rather, they make the voices less frightening. In other words, antipsychotics are not magic bullets that selectively target psychotic symptoms. They are blunt tools that cause a cascade of effects on multiple neurotransmitter systems. An important draw-back of these drugs is that they produce considerable sedation, often leaving patients feeling lethargic and debilitated.

Heroin also produces a range of neurobiological actions, including some leading to sedation. But, unlike antipsychotics, heroin engen-ders many positive effects on mood, including remarkable feelings of well-being. So, yeah, I can see how heroin might be more effective than many antipsychotic medications are at quieting the voices in the heads of some patients suffering from psychosis. I also can see how the drug might be more reinforcing. If patients like it, then they are more likely to take it. By contrast, most patients don't like taking the traditional antipsychotic medications.

After spending an entire day with Barbara, I was convinced that I needed to learn more about heroin. I needed to learn more from her. For example, I still didn't quite understand why a person would even consider using heroin, especially in the face of the apparently considerable risks. I wanted to know more. I felt extremely uncom-fortable, too, being so ignorant about a topic on which I was sup-posedly an expert. Our conversation lit a fire in me. I was determined to take steps to remedy my ignorance.

Fortunately, I was due for a sabbatical the following year. Barbara suggested that I spend some of my leave in Geneva working in the heroin clinic. This would allow us to continue our interactions while I learned firsthand about heroin in a clinical setting. I jumped at the opportunity.

In 2015, I spent several months in the Geneva clinic where Bar-bara treated heroin-dependent patients. Early on, I still had to deal with some of my most stubborn prejudices about the drug and the people who use it. I thought most of them acquired their addiction as a result of having been prescribed an opioid medication to deal with some other ailment. I was wrong. Despite the current false narrative, the addiction rate among people prescribed opioids for pain in the United States, for example, ranges from less than 1 percent to 8 percent.[1]

I now know, too, that most heroin users do not become addicted to the drug.[2] Your chances of becoming addicted increase if you are young, unemployed, and/or have co-occurring psychiatric disorders.[3] That is why the Swiss ensure that all heroin patients have a social worker, a psychologist, a psychiatrist, and other health professionals as part of their treatment team.[4] Not only are medical and mental-health issues addressed, but crucial social services are provided. All patients have housing, and many are employed.

Other myths I held were shattered each day I spent in the clinic. For instance, patients were required to show up at scheduled times twice a day—once in the morning and once in the evening—seven days a week. Like a Swiss watch, so-called junkies were reliably on time. They were almost never late. And as a result of being in the program, their health improved; they were happy and living responsible lives. It became impossible for me to retain the misguided notion that heroin addicts are irresponsible degenerates.

The Swiss introduced this form of treatment in the 1990s, when there was grave concern about the spread of the HIV virus through contaminated needles used to inject street heroin. In response to this growing health crisis, which was left unchecked by the typical law-enforcement approach that emphasizes drug-supply reduction, the Swiss government implemented the pragmatic approach of providing a select group of addicts with heroin, clean needles, and other services as part of their treatment. The approach worked. People stayed in treatment. The number of new blood-borne infections, such as HIV and hepatitis C, dramatically decreased. Petty crimes committed by heroin users also went down. And no heroin user has ever died while receiving heroin in the clinic.

I don't want to leave you with the impression that heroin maintenance is a panacea. It's not. It's not even a *cure* for heroin addiction; it's simply a *treatment*. There are no cures in psychiatric medicine. We don't have a cure for depression, nor do we have a cure for schizophrenia or anxiety. We merely have medications and therapies that treat symptoms, and this allows patients to function better, despite their illnesses.

But what is also true is that the psychosocial disruptions that led to a diagnosis of heroin addiction are no longer present in most patients enrolled in heroin-maintenance programs. These heroin users are healthier and more responsible. According to the *DSM-5*, this doesn't matter. These individuals are still saddled with the label "opioid use disorder"; they are still considered addicts. Except now they are described as addicts in remission. Stated another way, once an addict, always an addict. This classification isn't unique to opioid addiction. It also applies to other drugs, including alcohol, amphetamine, cannabis, and cocaine.

There is no scientific basis for the continued labeling of these individuals as addicts. This life sentence seems to be based purely on anecdote and convention. It is my hope that this diagnostic travesty is corrected in future editions of the *DSM*.

This in no way detracts from the success of heroin-assisted programs in Switzerland and other places. Some twenty years after the Swiss implemented their programs, several other European countries, including Belgium, the Netherlands, Germany, and Denmark, now employ similar approaches to treating heroin addiction in those who have failed repeatedly in conventional treatment programs, including abstinence- and methadone-oriented ones. Patients in these programs, like those in Switzerland, hold jobs, pay taxes, and live long, healthy, and productive lives.

Heroin, it turns out, is an effective treatment for heroin addiction. This was big news to me in 2015. I was excited about going back to the United States and sharing what I had learned. At the time, we were being inundated with a continuous stream of dire warnings attesting to the ravages caused by opioids. We were also routinely told that treatment was ineffective or inadequate.

When I returned from Switzerland, I agreed to participate in a panel discussion on "Heroin: A National Epidemic" (the organizers' words). I almost never take part in panel discussions about drugs because invariably at least one fellow panelist will present erroneous information, but if I correct the person, then I look like a jerk. I agreed to participate this time because I wanted to share what I had learned in Geneva and because the timing was perfect.

Peter Shumlin, then the governor of Vermont, was also a participant. In 2014, as I've mentioned, he gained national recognition in response to his focusing entirely on the "heroin crisis" in his State of the State address. He urged Vermonters to view addiction as a health problem and not a criminal-justice issue. The press loved it. Shumlin was seen as progressive and forward-thinking.

He wasn't. It quickly became clear that he was just another drug-stupid politician. During our discussion, I shared my Geneva experience and explained the treatment successes of several countries. I proposed that we should offer this treatment here in the United States. Shumlin, in a dismissively arrogant tone, responded by saying something like, "Americans don't need to take cues from any other nation." I couldn't believe it. His comment made me angry and left me dispirited. It is precisely this type of willful ignorance that prevents so many people from receiving treatment that actually works, whether it involves heroin maintenance or some other form of treatment.

Unfortunately, many Americans share Shumlin's view. Perhaps even more troubling is the large number of U.S. physicians and scientists specializing in drug addiction who find the idea of providing heroin to heroin-addicted patients simply wrong. For such physicians and scientists, it doesn't matter how well this treatment works. Or that it has been scientifically validated. Heroin use, even as a treatment, just feels immoral.

This ideological rigidity is one of the main reasons that heroin-assisted treatment is rarely mentioned in the United States. Heroin maintenance isn't discussed as a possible treatment option, nor is it taught as part of the medical education that budding physicians and addiction experts must complete. This seems a dereliction of duty.

I sometimes reflect longingly on the discussions I had with Barbara and her team about addiction, opioids, and life in general. Our conversations weren't constrained by puritanical notions that shame people into constricted and insipid thoughts, expressions, and lives. It was as if the shackles had been taken off my thinking, especially about drugs. I recall a talk that I had with Anne, one of the team's physicians. In response to my lament that physicians in the United States will probably never consider heroin a viable treatment option, for any condition, Anne said something I won't soon forget. First, she told me about how her patients often describe heroin in such loving terms. Then, with her eyes fixed on mine, she asked me, "How can I be against love?"

These experiences changed me. They made me question everything I thought I knew about heroin. I no longer believed much of the nonsense I had been taught about the drug. Nor did I now believe, as I once did, that heroin inevitably leads to death or some other tragic end. All the evidence from research clearly shows that most heroin users are people who use the drug without problems, such as addiction; they are conscientious and upstanding citizens.

I recognize that this statement requires some defense.

Since the early 1900s, when heroin was outlawed in the United States, the drug and its users have been denigrated in the popular press, in politics, in art, everywhere. Sure, periodically there have also been sympathetic portrayals of certain heroin users, especially if they were white, young, and physically attractive. But the number of these more compassionate depictions pales in comparison to those that are negative.

Newspaper headlines routinely read like hysterical heroin-danger warnings: HEROIN SUSPECTED IN 20 DEATHS IN 2 WEEKS[5]. Artists regularly produce influential works that cement heroin's evil reputation. Who among us wasn't deeply moved when Johnny Cash sang "Hurt," poignantly describing the horrors of heroin use? Or when Neil Young sang "The Needle and Damage Done," a song inspired by the death of his former bandmate Danny Whitten, a heroin user?

You might not know this, but Johnny Cash never used heroin, nor did he receive any special training on its effects. So his view of the subject is probably not the most authoritative. Trent Reznor, the front man for Nine Inch Nails, wrote "Hurt," and he definitely used heroin. But his use alone doesn't make him an expert either. Equally important, even before Reznor used heroin, he suffered from depression. As I said earlier, people diagnosed with a psychiatric disorder have an increased chance of becoming addicted. This makes it extremely difficult to disentangle the effects of depression from those of heroin when trying to determine the actual cause of Reznor's misery when he penned "Hurt." Likewise, Danny Whitten, the inspiration for Neil Young's song, didn't die from a heroin overdose; he died from an overdose of prescription sedatives and alcohol.

The point is that a person shouldn't be considered an expert on heroin merely because she wrote a song, or an article, about the horrors of the drug. Nor should a person be deemed an authority on the drug simply because he used it in a pathological manner. It's like saying that Donald Trump is a gynecologist because he once had a morbid predilection for grabbing women by the crotch without their permission. Moreover, most popular depictions of heroin use are less than accurate and do not tell the whole story. All one has to do is dig a little, and it becomes abundantly clear that heroin—or any other opioid, for that matter—isn't the villain it has been made out to be.

When it comes to street heroin, far more concern should be focused on contaminants that may be contained in the substance. Today, illicit heroin is frequently adulterated with stronger opioids like fentanyl and its analogs. The adulterants are often much more dangerous than the heroin itself. These substances produce a heroin-like high but are considerably more potent. This, of course, can be problematic—even fatal—for users who ingest too much of the substance thinking that it is heroin alone or another single opioid. Fentanyl is blamed for the death of rock star Prince. It has been reported that he died after taking a fentanyl-containing pill he thought was Vicodin.

One obvious solution to many of these accidental deaths is to legalize heroin, as we have done with marijuana in eleven states and with alcohol. Legalization would ensure a minimum level of quality control. During Prohibition, alcohol produced in illicit stills frequently contained contaminants that made people sick or even killed them. This problem went away when Prohibition was repealed. In the meantime, we should offer free, anonymous drug-purity testing services. If a sample contains adulterants, users would be informed. As I noted in earlier chapters, these services already exist outside the United States in places such as Austria, Belgium, the Netherlands, Portugal, Spain, and Switzerland, where the first goal is to keep users safe.

In an effort to circumvent the unpredictable nature of illicit heroin markets, some people obtain prescriptions for opioids as substitutes. On the one hand, this is a good thing because the purity of street heroin is frequently poor. Prescription opioids are usually higher quality and pharmaceutical grade. But, on the other hand, popular prescription medications such as Percocet, Vicodin, and Tylenol 3 contain an extremely small dose of opioid in combination with a considerably larger dose of acetaminophen (a.k.a. paracetamol)—and excessive acetaminophen exposure is the number one cause of liver damage in the United States.[6] Some users may unwittingly risk liver damage by taking too many of these pills.[7] We need to inform people not to overdo it on opioid pills that contain acetaminophen because it can be much more fatal than the low dose of opioid usually contained in these formulations.

THE SAD FACT remains that far too many people are victims of preventable opioid-related deaths. Too frequently, I receive emails, letters, phone calls, and visits from parents who recently lost a child from what was described to them as an opioid overdose. My interactions with these grieving parents are heart-wrenching. As a parent myself, I can't imagine ever recovering from such a loss. I feel a profound sense of compassion for them, and I provide whatever help I can to these broken parents.

I remember meeting Tatianna Paulino after the death of her son Steven Rodriguez. Steven, who was also known as A$AP Yams, was the founder of the hip-hop collective A$AP Mob. In the early hours of January 18, 2015, his mother learned of his death as she raced down a Bronx highway headed toward the Brooklyn hospital that received his body. Steven had succumbed to an apparent opioid overdose. This seemed consistent with the fact that he had gone to drug treatment once before. It also was consistent with the stories told to the press by some of his friends. A$AP Rocky, for example, said in an interview with *The New York Times* that Steven "always had a struggle with drugs. That was his thing."

Steven's reported drug of choice was "lean," a.k.a. purple drank, or syrup. The concoction is a mixture of a flavored soft drink with cough syrup containing codeine and promethazine. The opioid codeine is one of the naturally occurring chemicals found in the opium poppy. In medicine, codeine is used as a cough suppressant as well as a pain reliever. Some people also use codeine to get high because it can relieve stress and produce sedation with mild euphoria. Opioids, including codeine, also stimulate the release of histamine, which can lead to itchiness, nausea, vomiting, and other unpleasant symptoms.

Promethazine is an antihistamine used to treat allergy symptoms, such as itching, runny nose, sneezing, itchy or watery eyes, hives, and itchy skin rashes; it's also used to reduce nausea, vomiting, and insomnia. In practice, promethazine, as well as other antihistamines, can eliminate many of the negative side effects produced by opioids. But, I suspect, the primary reason for its inclusion in the lean drink is its pronounced sedating properties.

Tatianna knew her son suffered from sleep apnea and wondered if it played a role in his demise. She was concerned that the portrayal of her son's death as a "drug problem" seemed too simple, too convenient. It didn't explain the actual cause of his death. Nor was it an accurate enough explanation to ensure that Steven's experience could be used to prevent something similar from happening to someone else's child. A lot more seemed to be going on here. Sure enough, Tatianna learned that her son was under considerable pressure to produce hit recordings. He was also broke, and, to add further insult to injury, he was being squeezed out of the group he had put together. Given these stressful circumstances, it's not difficult to see how Steven's use of opioids might have been a way for him to decompress. Opioids certainly are well-suited for this task.

The problem is, however, that most lean concoctions contain both an opioid and an antihistamine. If a nontolerant person takes a large dose of an opioid combined with an antihistamine, especially an older one, such as promethazine, which is heavily sedating, that person's chances of experiencing fatal respiratory depression increase dramatically. The evening before his death, Steven drank lean. What's more, his toxicology report revealed that he had also taken oxycodone (an opioid) and alprazolam (a benzodiazepine a.k.a. Xanax). Both are also known for their anxiety-relieving and euphorigenic effects. But when they are combined with other sedatives, especially at large doses, these drugs can become deadly.

I can't say whether Steven was aware of the potential dangers of mixing opioids with other sedatives. If he wasn't, he certainly wasn't alone. Several celebrities have died as a result of combining opioids with other sedatives: DJ Screw, Pimp C, Heath Ledger, Cory Monteith, Philip Seymour Hoffman, and Tom Petty, among others. Headlines announcing these deaths almost always single out the opioid as the killer, which is not only less than accurate but also incredibly irresponsible.

It is certainly possible to die from an overdose of an opioid alone, but such overdoses account for a minority of the thousands of opioid-related deaths. Most are caused when people combine an opioid with alcohol, an anticonvulsant, an antihistamine, a benzodiazepine, or another sedative. People are not dying because of opioids; they are dying because of ignorance.

Explaining these facts to Tatianna was difficult because I felt for her. I also felt fortunate to have had the opportunity to get to know her and to learn more about her son. I remember her telling me that Steven could light up a room with his smile, that he was delightfully mischievous, that his entire body shook when he laughed. She also met my son, Damon, who was only a few years younger than Steven was. She told me how lucky I was to have him. In one of our meetings, she turned to me and sadly said that she wished public-health messages would simply state, "Don't combine opioids with other sedatives." "Perhaps," she continued, "my son would be alive today." I was heartbroken and speechless. I couldn't imagine what I would do if the shoe were on the other foot.

Her simple message was spot-on. I was determined to help spread it. I grew increasingly frustrated with politicians who overstate the harms of opioids alone. This merely diverts attention and resources away from the real concerns and decreases our ability to take the most appropriate steps to keep people healthy and safe. The undeniable fact is: Opioids have been used safely for centuries. They've been used to lessen people's misery and as important instruments in the physician's toolbox. We, as a society, should recognize that people will always use these drugs, whether authority figures like it or not.

The first time I did heroin I was well over forty. It wasn't a youthful indiscretion, as many politicians disingenuously claim about their own drug use. It was deliberate. It was also unremarkable. My friend Kristen asked if I would be interested in trying heroin with her. She had never done it but wanted to try it. Same here. So one Friday evening, we did. Unlike in the movies, we didn't use needles. (By the way, nor do most heroin users.) We each snorted a short, thin line. Immediately, we detected the nice, characteristic opioid effects, including a dreamy light sedation, free of stress. We talked, reminisced, laughed, exchanged ideas, and carefully documented our drug effects. After they had worn off, we called it an evening and went home.

I was struck by how inconsistent my experience was compared with the chaotic heroin-use scenes typically portrayed in the popular culture. This reinforced my belief that horrible outcomes are too often misattributed to heroin. I no longer feared the drug or pictured a ruinous outcome if I took it. I was now a heroin user. In fact, heroin is probably my favorite drug, at least at the moment.

But, to be clear, I am not an addict, and I don't say that to distance myself from those who may struggle with heroin addiction. It's just a fact. I don't jones for the drug, nor do I use it daily. In fact, the frequency of my heroin use is about as occasional as that of my alcohol use. I have never failed to meet my obligations as a result of the drug or its effects, nor have I involuntarily experienced symptoms that would suggest that I have a problem. I don't bang (not that there is anything wrong with injecting), nor do I have tracks. I have never nodded off or slurred my speech after taking the drug. No one could tell that I am a heroin user simply by looking at me. The same is true for most other heroin users.

My heroin use is as rational as my alcohol use. Like vacation, sex, and the arts, heroin is one of the tools that I use to maintain my work-life balance. All our lives are filled with pain, stress, and heart-break. In order for me to remain relatively intact psychologically and to be a humane person, I have developed successful strategies to mitigate the inevitable harm caused by difficult people, impossible situations, unrealistic expectations, and myriad other life stressors. But, to be absolutely clear, I also enjoy heroin for the mere pleasure of its effects.

A few years ago, I was asked to do a three-year bid as department chair. I was honored to have been asked but wanted to consult with others before deciding. Many of my wiser friends and colleagues—some who had previously served in similar administrative roles themselves—advised me not to accept the position. They feared I'd get bogged down in petty departmental politics and be waylaid from my own work. Several echoed the sentiment often attributed to Henry Kissinger: "Academic politics are so vicious precisely because the stakes are so small."

In the end, though, I agreed to serve. I wanted to help shape the future mission of our department. I wanted to make sure we were doing our share to include, as part of our faculty and student body, individuals from groups that had been and continue to be shut out of elite institutions. I also wanted to give back to a department that had been so generous and supportive of me and my work. My service as department head would be my way of saying thank you.

During my tenure as chair, my gratitude was eroded. For example, I sadly learned that it would be difficult, if not impossible, to increase the number of black faculty members beyond token levels. Black candidates, it seemed, not only had to have an extraordinary academic record but also had to be deemed nonthreatening. If a current faculty member felt threatened by the candidate's independence, intellect, popularity, success, whatever, it was a wrap. The applicant didn't stand a chance. Of course, "nonthreatening" is a vague and capricious factor that is never explicitly stated during hiring discussions. Instead of focusing on the applicant's record, these meetings too often descended into innuendo and whispering campaigns based on rumors from anonymous third-party sources. The anonymous information is usually disclosed by faculty members who are the biggest proponents of "diversity."

In the university setting, the term *diversity* has replaced the spirit of redress and has come to represent anybody from black faculty to military veterans. Well, I am both, but have yet to be subjected to discrimination because I'm a veteran. I now cringe whenever I hear colleagues going on about the importance of having a diverse campus community.

Operating within this context caused me a great deal of cognitive dissonance, especially because a large part of my job as department chair consisted in advocating for our faculty. Sometimes it meant presenting before a university committee a persuasive tenure or promotion case on behalf of a colleague. Other times it meant negotiating to secure a coveted university apartment for her or helping to guarantee a spot for his child in Columbia's K–8 primary school. I frequently wondered, "How can I continue to serve on behalf of people who actively undermine my just efforts?" It was dispiriting.

One of my favorite ways to unwind and rejuvenate is to watch live comedy shows. It helps me not to take myself too seriously. I love laughing, especially laughing at myself. I'm reminded that I, too, am fallible and flawed. As a result, I try to be more understanding and forgiving of others, even if they disappoint me. Comedy has helped me to be a better person.

So has heroin. There aren't many things in life that I enjoy more than a few lines by the fireplace at the end of the day. Billie Holiday's soul-stirring voice sets the scene and mood: "God bless the child that's got his own." Holiday herself was an avid heroin user. She was criticized, of course. Her response, according to biographer Farah Jasmine Griffin, was that neither of her parents used drugs and she outlived them both: "Heroin not only kept me alive—maybe it also kept me from killing."[8] I know the feeling.

In these serene moments, I reflect on my day, hoping that I wasn't the source of anyone's anguish. I replay interpersonal interactions with the goal of seeing things from the other person's point of view. I am acutely aware of my role and responsibilities, recognizing that my interactions with others, especially subordinates, can cause anxiety or bruised feelings that negatively impact the individual's subsequent interactions with their loved ones or with others.

Heroin allows me to suspend the perpetual preparation for battle that goes on in my head. I am frequently in a state of hypervigilance in an effort to prevent or minimize the damage caused by daily living in my own skin. When heroin binds to mu (μ) opioid receptors in my brain, I "lay down my burden" as well as "my sword and shield," just as described in the Negro spiritual "Down by the Riverside."

The world is alright with me. I'm good. I'm refreshed. I'm prepared to face another day, another faculty meeting or obligatory function. All parties benefit.

I recognize that my experience with heroin wildly conflicts with depictions of the drug as causing users to become emotionally numb. Certainly, extremely large doses of the drug can produce this effect; they can even render a person unconscious. But such effects are virtually nonexistent for and are definitely unwanted by most who seek to enjoy heroin-related effects. Statements attesting to the "numbness" caused by heroin are gross mischaracterizations. They reduce heroin's effects to something like a deprivation of feelings. It is, precisely, the feelings—forgiving, open, and tranquil—produced by the drug that inspire me to be a more empathetic person. In other words, heroin *enhances* my ability to feel.

Also, it's important for me to make clear that my use of heroin—or any other drug—isn't usually a solo pastime. Regularly, some of my closest friends and I bond over the sweet, earthy smell of burning opium.

"I spoke some foolish things to her." Fabrice made me laugh as he ran down a blow-by-blow account of an embarrassing night he spent with a mutual friend in Paris. He had had too much alcohol and now wished he could take back his words. He can't.

Fabrice and I were in a Prague hotel. We had been invited to give presentations at a congress on drugs. Neither one of us was particularly excited about the prospect of speaking to this audience. It was largely composed of psychiatrists who refer to themselves as addictionologists. If ever there was a group of individuals resistant or immune to solid evidence inconsistent with their own worldview, this was the group.

I had just arrived from New York City after a nine-hour flight. I was exhausted. In addition, during the two weeks prior, I had given talks in Los Angeles, Lubbock, and Boston. I did this while still teaching my twice-weekly course at Columbia and my Friday night course at Sing Sing. By now, I was suffering from a respiratory infection and felt mildly feverish and achy. I also had a persistent cough that made my scratchy, sore throat feel worse.

It had been nearly a year since I last saw Fabrice. I'd missed him. No matter how much time passes between our meetings, we always pick right back up from our previous time together, without awkwardness or weirdness. Fabrice is family.

Sitting in that unremarkable hotel room, we smoked opium and laughed almost nonstop. We told stories that highlighted faux pas made by each of us. We made plans to see each other more frequently, despite residing in different countries. We exchanged new information we had learned from our research and about our drugs of choice. We inquired about each other's families and plotted a holiday together soon.

The hours flew by. Thankfully, opium had lessened my symptoms just in time for dinner, where we met with other congress presenters. Many of them enjoyed wine or another alcoholic beverage with their meal. This no doubt relaxed them and facilitated their own social interactions. Fabrice and I were already there. It turned out to be a lovely evening, and the congress wasn't bad either.

My ongoing experience with opioids continually forced me to update my thinking. Shortly after completing my PhD, a former professor suggested that I watch the 1996 film *Trainspotting*. This professor implied that I'd learn something important about heroin withdrawal. I watched it, and as a result, I thought I was informed.

In 2017, I rewatched it. This time, I cringed, especially during the withdrawal scenes. It was too sensationalistic and corny, and by portraying opioid withdrawal as a near-death experience, the film reinforced incorrect and harmful stereotypes about the drug and its users. I knew this depiction wasn't representative of most users' experience, because by now I had been through mild heroin withdrawal on more than one occasion. Never had I been terrorized by visual hallucinations during the detox. Never had I experienced the agonizing pain that purportedly drives users to do anything to get another hit. I had experienced absolutely none of that *Trainspotting* bullshit.

Granted, in the past, I had only used low intranasal doses of heroin for no more than about ten consecutive days at a time. Nonetheless, this pattern of use was enough to produce some withdrawal symptoms when I abruptly stopped using. The symptoms would start about twelve to sixteen hours after the last dose. At most, I felt like I had a case of the twenty-four-hour flu: chills, runny nose, nausea, vomiting, diarrhea, and some mild aches and pains. The bottom line is that it was unpleasant but certainly not dramatic or life-threatening.

Still, I had to face the question of why media portrayals of heroin withdrawals were so inconsistent with what I knew from my own experience and from what I've read in the scientific literature. Perhaps the amount of heroin I was using was too low? Or perhaps I needed to use on more consecutive days? I also knew that many avid opioid users report that withdrawal from longer-acting opioids, like methadone, is far worse than it is from heroin. Taking these issues into account, I remembered that I had in my possession a large pill bottle of extended-release morphine. The pills once belonged to a relative who had been prescribed them for pain prior to dying. It didn't seem right to let the pills go to waste.

So, as part of my experiment, I began taking daily oral doses of morphine, about 30 to 45 mg, and continued for approximately three consecutive weeks. I also used heroin during this period. I planned my "quit day" to be about forty-eight hours prior to a major talk I was scheduled to give. In this way, I would have at least an entire evening to deal with the symptoms. I was going to prove, once and for all, at least to myself, that withdrawal was an inconvenience that could be dealt with without failing to meet major obligations.

THE EVENING OF WITHDRAWAL

It was shortly before midnight, but I couldn't sleep. I was experiencing one of the worst withdrawal symptoms I had ever felt. It wasn't the nausea, vomiting, or even diarrhea. Each had subsided. Besides, there was no more food to get rid of. It wasn't my cravings for the drug. I could take it or leave it. What I really wanted was sleep.

But that wasn't in the cards. The pain in my abdomen was too intense for me to be able to drift off. It was agonizing and unremitting. It was a pain that I would not easily forget. It was a new pain, unlike any I had previously experienced. It was so intense that it radiated throughout my entire body. The light touch of Robin's hands on my leg or arm to soothe me merely exacerbated the pain. It lasted for hours, and nothing seemed to alleviate it. We tried aspirin and ibuprofen; neither worked. Cannabis and triazolam—both failed, too. Triazolam is a benzodiazepine and is used to treat insomnia. I wasn't afraid to take it in this situation because I had only a small amount of opioid in my system.

At this point, Robin hoped I'd go to the emergency room. She gently floated the idea. The distress plastered across her face, however, told a different story, a more urgent one. She was worried, deeply worried. And this worried me. It's true, my abdominal pain was absolutely dreadful. But I knew that it wasn't life-threatening and that it would eventually lessen. Robin didn't know this though.

I had to do something. Act fast. I crushed two 0.25 mg triazolam tablets and snorted them. I knew the drug would reach my brain quicker if I snorted the pills than if I swallowed them. I also knew that two pills would definitely knock me out. Within fifteen minutes, I was sound asleep for the next six hours. Robin, vigilantly monitoring my status throughout, was relieved.

When I woke, the abdominal pain was still present but not remotely as intense. Minor flu-like symptoms, including a runny nose and slight queasiness, also remained. None of this particularly bothered me. I was just relieved that it played out as I thought it would and that it was over. Well, almost over. I shifted my attention to preparing for my upcoming talk, which was going to take place in less than two hours.

After the host introduced me, I started a talk titled "Everything You Thought You Knew about the Opioid Crisis Is Wrong." Then I told the audience that my talk came at a fortuitous time because I was in the midst of opioid withdrawal. They all laughed, of course; no one seemed to believe me. By all accounts, the talk went quite well. The room was full. Attendees appeared engaged and stayed put until the end. Relevant questions and comments followed. In sum, the talk concluded without a hitch.

Going through opioid withdrawal wasn't a particularly pleasant experience. And I don't have plans to do it again anytime soon. But I am glad to have done it. This experience confirmed a few things I already knew. First, opioid withdrawal is not life-threatening. The same cannot be said for alcohol withdrawal. You will not read within these pages that I conducted an alcohol withdrawal self-study. Second, withdrawal symptoms do not equate to addiction. Despite the fact that I underwent opioid withdrawal, I have never met criteria for opioid addiction. Similarly, we would not label a person an addict on the mere basis of experiencing withdrawal symptoms after abruptly discontinuing an antidepressant. Finally, media portrayals of opioids focus almost entirely on negative outcomes—and even these are often exaggerated. This situation led me to act and speak out. I want to help people see through the "opioid crisis" hysteria and all the damage it causes. I also want to ensure that others are afforded safe opportunities to benefit from the serene bliss opioids can offer, should they so choose.

The Journey

You cannot know what you will discover on the journey,
what you will do with what you find,
or what you find will do to you.

James Baldwin

On October 4, 2019, I gave a keynote address in Tulsa, Oklahoma, at the Zarrow Mental Health Symposium. It was my first visit to the state. The room was packed with an audience of more than five hundred mental-health professionals and clients. They'd come to hear me give a lecture titled "Drug Talk for Grown-Ups." I didn't know what to expect, especially from this crowd. How would they respond when they realized I wasn't there to scapegoat drugs? How would they respond to my conclusion that recreational drugs should be legally regulated and available for adult use?

I began by explaining that my career had taken me on an intellectual and geographical journey. I had conducted dozens of laboratory studies investigating the behavioral and neurobiological effects of psychoactive drugs and communicated my findings in respected scientific journals. I had traveled from New York City to Accra, Ghana; to Salvador, Brazil; to Nassau, Bahamas; to Edmonton, Canada; to Chiang Mai, Thailand; to Tel Aviv, Israel; to Oslo, Norway. In fact, I had spent time on five continents and in innumerable locales within each as part of my research and continuing education. It had been one hell of a journey.

MY FINDINGS

I discovered that the predominant effects produced by the drugs discussed in this book are positive. It didn't matter whether the drug in question was cannabis, cocaine, heroin, methamphetamine, or psilocybin. Overwhelmingly, consumers expressed feeling more altruistic, empathetic, euphoric, focused, grateful, and tranquil. They also experienced enhanced social interactions, a greater sense of purpose and meaning, and increased sexual intimacy and performance. This constellation of findings challenged my original beliefs about drugs and their effects. I had been indoctrinated to be biased toward the negative effects of drug use. But over the past two-plus decades, I had gained a deeper, more nuanced understanding.

Sure, negative effects were also possible outcomes. But they represented a minority of effects; they were predictable and readily mitigated. For example, the type of drug use described in this book should be limited to healthy, responsible adults. These individuals fulfill their responsibilities as citizens, parents, partners, and professionals. They eat healthy, exercise regularly, and get sufficient amounts of sleep. They take steps to alleviate chronic excessive stress levels. These practices ensure physical fitness and considerably reduce the likelihood of experiencing adverse effects. Equally important, I learned that people undergoing acute crises and those afflicted with psychiatric illnesses should probably avoid drug use because they may be at greater risk of experiencing unwanted effects.

The vast amount of predictably favorable drug effects intrigued me, so much so that I expanded my own drug use to take advantage of the wide array of beneficial outcomes specific drugs can offer. To put this in personal terms, my position as department chairman (from 2016 to 2019) was far more detrimental to my health than my drug use ever was. Frequently, the demands of the job led to irregular exercise and poor eating and sleeping habits, which contributed to pathological stress levels. This wasn't good for my mental or physical health. My drug use, however, has never been as disruptive or as problematic. It has, in fact, been largely protective against the negative health consequences of negotiating pathology-producing environments.

I am not alone. A large number of people use government-prohibited substances for similar reasons. In the United States, a recent nationwide survey revealed that thirty-two million Americans had used at least one such drug in the past month.[1] Contrary to popular media portrayals, most drug users are not addicts. They are responsible members of their communities. They pay their bills and taxes on time; they take care of their families; and they volunteer in their local and global communities. They are artists, engineers, firemen, homemakers, judges, lawyers, pastors, physicians, politicians, professors, schoolteachers, scientists, social workers, truck drivers, writers, and many other types of professionals.

But most go to great lengths to conceal their drug use, frequently living double lives. The toll this takes on an individual, of course, can vary widely, depending upon personal attributes and societal norms. Some people experience tremendous anguish due to their duplicity, while others assuage their guilt with self-satisfying rationalizations.

Regardless, it's not difficult to understand why so many people remain in the closet about their drug use. For the past hundred-plus years, communities around the world have been inundated with information that almost exclusively emphasizes the harmful, even deadly, effects produced by nearly all the drugs discussed here. Drug users are routinely vilified and imprisoned, and sometimes killed, for merely being identified as such. Even when respected, sober-minded experts raise doubt about the veracity of drug-blaming and exaggerated claims, efforts to ban the targeted drug and to excoriate specific users and sellers proceed with little fruitful resistance.

In the late nineteenth century, alcohol and drinkers were the targets in the United States. It was asserted that the drug "takes the kind, loving husband and father, smothers every spark of love in his bosom, and transforms him into a heartless wretch, and makes him steal the shoes from his starving babe's feet to find the price for a glass of liquor. It takes your sweet innocent daughter, robs her of her virtue and transforms her into a brazen, wanton harlot."[2] These negative narratives became so plentiful that Congress was persuaded to amend the Constitution, banning the manufacture, sale, or transportation of alcoholic beverages. The Eighteenth Amendment took effect on January 17, 1920. It would take almost a decade and a half—and the belief that alcohol-tax revenue would lower income taxes—before reason prevailed. On December 5, 1933, the Twenty-

First Amendment repealed the Eighteenth Amendment, making it the only one ever to be repealed.

Today, one hundred years later, nearly identical bankrupt arguments are hawked to support bans on other drugs in several countries, the United States included. Judging from the dominant response to the current North American opioid situation—increased restrictions placed on the legal availability of these drugs—little has been learned from the alcohol-prohibition experience. As had occurred during the prohibition era, loads of people still consume so-called banned drugs, including opioids, cocaine, and psychedelics. Many of these people are forced to obtain their drugs of choice from illicit, unregulated markets, where there aren't any quality controls. Thus, just as during Prohibition, thousands of people have died from ingesting drugs contaminated with poisons, impurities, and other unknown substances.

Alcohol tainted with large amounts of methanol killed thousands of drinkers and left many others blind during Prohibition. As Deborah Blum masterfully explains in her authoritative work, *The Poisoner's Handbook*, the U.S. government callously caused many of these deaths.[3] Even before Prohibition, as early as 1906, federal officials required producers of industrial alcohol—used in antiseptics, medicines, and solvents—to add methanol and other chemicals to their batches so their products would be undrinkable. This policy was implemented to deal with manufacturers who sought to avoid paying taxes on potable alcohol. The Prohibition era brought with it sophisticated traffickers who obtained industrial alcohol, redistilled it to be quaffable, and sold it to the public and speakeasies. Government authorities were not pleased. Alcohol had been banned, but people continued to imbibe.

By the mid 1920s, the feds were fed up. They ordered industrial alcohol makers to add even more methanol—up to 10 percent—to their products, which proved to be particularly lethal. Illicit dealers were caught off guard, and redistilling industrial alcohol required much more effort. Most individuals, certainly most drinkers, were unaware of these developments. People continued to drink, and the alcohol-poisoning death toll continued to climb. By the time Prohibition ended, hundreds of thousands of people had been maimed or killed due to drinking tainted alcohol. An estimated ten thousand of these individuals died as a result of the government alcohol-poisoning program. Neither accumulating deaths nor public outcry compelled the government to change its deadly alcohol-poisoning policy. This war-on-alcohol tactic remained in effect until Prohibition was repealed.

Thinking about these events, I can't help but see the hypocrisy of our current approach that allows the government to prosecute as a murderer anyone who provided the drug to a fatal-overdose victim. The fact of the matter is that many dealers, especially the low-level ones, don't know the complete composition of the substances they sell. It is true that some drugs sold by these individuals may contain harmful adulterants. But unlike Prohibition authorities, their intent certainly isn't to kill or harm consumers. If our current government—or any government—were genuinely concerned about the health and safety of drug users, it would ensure that free, anonymous drug-safety testing services were widely available. This practical approach informs users of the contents of their substances and decreases the likelihood of people ingesting fatal amounts of unknown substances.[4]

The parallels between the government-mandated methanol-poisoning policy and the current practice of combining an opioid

with acetaminophen in a single pill are frightening. Several pharmaceutical companies offer such FDA-approved products. The pain
medication Percocet, for example, contains a low dose of the opioid
oxycodone and a much larger amount of acetaminophen. It is claimed
that such formulations provide complementary, more effective pain
relief than does the opioid alone. Even if this is true, which I do
not concede, the risk-to-benefit ratio is simply not favorable when
one considers the potential lethality and toxicity of acetaminophen.
Acetaminophen-induced liver toxicity is the most common cause of
acute liver failure, which can be fatal. Approximately 6 to 10 g of
acetaminophen taken for two consecutive days is enough to cause
liver damage.[5]

The typical prescribed dose of Percocet contains 325 mg of acetaminophen and only 5 mg of oxycodone. That means twenty pills
per day taken for multiple days can yield enough acetaminophen to
produce liver toxicity. The same number of pills, however, only provides a relatively small amount of oxycodone (100 mg) for the experienced opioid user. Many consumers of opioid pain relievers don't
even know that these medications often contain acetaminophen.[6]
For me, the solution is simple: remove acetaminophen from opioid
pain formulations. The risks far outweigh the benefits.

I have also observed that the enforcement of drug laws, regardless of the country, is frequently carried out in a selective manner.
Individuals from despised and marginalized groups are disproportionately targeted, arrested, and imprisoned for drug-law violations,
even though recreational drug use is common in every strata of society. Overwhelmingly, the targeted are resource-poor people, and
their ability to obtain appropriate legal representation is practically
nonexistent. To add insult to injury, moralists and others reflexively
blame drugs for poor people's problems, including poverty. This ill-

conceived logic ignores the fact that most drug users are not poor and many have ample disposable income. Think about it. Drug trafficking is a multibillion-dollar industry. "Poor people alone can't sustain the operating budgets of drug cartels," my friend Raphael once said. We were discussing the Gordian knot that is Brazilian drug-war politics, on a temperate evening in his upscale Rio de Janeiro neighborhood. We were also enjoying some of Brazil's finest cocaine.

A CHANGED MAN

These observations forced me to take a long, uncomfortable look in the mirror. I had to acknowledge my own drug use. Like so many other privileged people, I had spent years hiding in the closet. Unlike the underprivileged, I wasn't subjected to humiliation, persecution, and death simply for being identified as a user. Perhaps I was protected *because* I remained in the closet? I don't know. What I do know, however, is that my conscience will no longer allow me to remain silent about my drug use, nor can I remain silent about the absurdity of punishing people for what they put into their own bodies. How could I? To this day, countless people are subjected to harsh punishments for using drugs. What kind of man would I be if I didn't publicly voice solidarity with these individuals? I'd be a hypocrite and a coward. I should know because I had been living as such for many years. I refuse to do so any longer.

I have been profoundly changed by my journey. I rediscovered the Declaration of Independence and the noble ideals expressed in it. It guarantees each of us "certain unalienable Rights," including "Life, Liberty and the pursuit of Happiness," so long as we don't infringe upon the rights of others. Simply put, it is my birthright to use substances in my pursuit of happiness. The point is that whether I use a drug or not is *my* decision; it is not the government's decision. Further, my responsible drug use should not be subjected to punishment by authorities. These ideas are central to our notions of liberty and personal freedom. The current punitive approach to dealing with recreational drug users is wholly un-American.

And it highlights the fact that our nation doesn't always live up to its virtuous ideals. This was particularly apparent during the slavery era. Numerous American heroes—such as Harriet Tubman, Nat Turner, and Henry David Thoreau—led rebellious acts to reconcile the government's actions with the country's promise of liberty for all its citizens. People such as Fannie Lou Hamer and Martin Luther King Jr. frequently quoted the Declaration in their efforts to eradicate racism. King said, in his *I Have a Dream* speech, "that all men, yes, black men as well as white men, would be guaranteed the unalienable rights of life, liberty, and the pursuit of happiness."

Yes, I know that reminding the nation of its promise, its foundational principles, has yet to create a perfect society—that would be a tall order—but by doing so, it can provide clear direction toward the ideals we should seek to achieve. I hope this book clearly shows that government bans on recreational drugs violate the spirit and promise of the nation's founding document.

The drugs described in this book should be regulated and legally available for adult consumption. We have already taken this approach with alcohol, tobacco, and, more recently, in a handful of states, with marijuana. The benefits are numerous. For starters, the legal availability of drugs fulfills the Declaration's promise of allowing responsible adults to pursue happiness as they see fit. In addition, a legally regulated drug scheme would create numerous jobs and generate hundreds of millions of dollars in annual tax revenue.

Also, such a scheme would markedly reduce drug-related deaths caused by accidental overdoses. A large proportion of these deaths are caused by adulterated substances purchased on the illicit market. A regulated market, with uniform quality standards, would virtually put an end to contaminated drug consumption and greatly reduce fatal, accidental drug overdoses.

"You are badass!" was the first comment I heard when I finished speaking to the Tulsa audience. A woman in her forties gleefully stood at the microphone when the floor opened for questions. She said that she shared my perspective but had previously remained silent because she feared she would be ridiculed. Other enthusiastic questioners expressed similar sentiments. Some wanted to know specific steps that could be taken in order to facilitate regulatory schemes that would permit adult drug use.

My responses were similar to many of the points I have emphasized within these pages. I reiterated the importance of requiring people to back up their claims about drugs with credible evidence. Too often anecdote alone and misinterpretation of data drive the unrealistic and inaccurate drug stories presented to the public. For example, despite the fact that there are practically no data indicating that recreational drug use causes a brain disease, many people, including some drug scientists, *believe* otherwise. But beliefs alone are insufficient to guide drug-education efforts and evidence-based health policies.

I also recommended that respectable middle-class drug users stop concealing their use. If more people followed this advice, it would be extremely difficult to pigeonhole all users as only irresponsible, troubled members of our society. I explained that my perspective had been heavily influence by King's *Letter from Birmingham Jail*, in which he made a compelling case for disobeying unjust laws.[7] I urged them to get out of the closet and to blatantly disregard laws that prohibit adult drug use, because such laws are ruthlessly unjust. I explained that I hope my writings and speeches inspire massive civil disobedience among the privileged class. I told them that they should stage mass protests whenever police tout "drug-crazed" myths to justify their excessive force.

Finally, I said that we have allowed our nation to build a huge law-enforcement apparatus in a misguided war-on-drugs effort. A Herculean task would be required to dismantle it, not to mention the loss of jobs for millions. So, instead, I suggested that we advocate for retraining and redirecting the efforts of this antidrug bureaucracy. Police don't receive any training in behavioral pharmacology. Yet we often require them to deal with drug-related issues and educate the public about what drugs do and don't do. Teaching police some basic information—such as that there are no drugs that create superhuman strength and that the effects of drugs are determined by the interaction between a user and her environment—would go a long way in dispelling the drug myths they often perpetuate. Moreover, the first priority of law enforcement should be to keep users safe, not to arrest them.

If the ideas expressed in this book are embraced, we can get on with the business of treating each other better and enjoying more meaningful and fulfilling lives. And isn't that what we all want?

Death Investigation Systems by State

State	Death Investigators	Elected or Appointed	Qualifications	Other requirements
Alabama	Coroner	Elected for four-year term	Complete twelve hours of training within six months of assuming office Complete twelve hours of in-service training on an annual basis	US citizen At least 25 years old Residence in the county of service Registered voter High school diploma
Alaska	Medical Examiner	Appointed	Licensed physician with expertise in forensic pathology	Licensed to practice in the state
Arizona	Medical Examiner	Appointed	Board-certified forensic pathologist	
Arkansas	Coroner	Elected for four-year term	Complete 16 hours of training	Residence in the county of service Registered voter
	Medical Examiner	Appointed	Licensed physician with expertise in pathology	Licensed to practice in the state
California	Coroner	Appointed or elected	49 of 58 counties, the elected sheriff is the coroner	
	Medical Examiner	Appointed	Licensed physician with expertise in pathology	Licensed to practice in the state

State	Death Investigators	Elected or Appointed	Qualifications	Other requirements
Colorado	Coroner	Elected for four-year term	Complete a basic coroner course Obtain certification in basic medical-legal death investigation within 12 months of assuming office Complete 16 hours of in-service training on an annual basis	US citizen Residence in the county of service High school diploma
Connecticut	Medical Examiner	Appointed	Licensed physician with expertise in pathology	US citizen Licensed to practice in the state Minimum of four years postgraduate training in pathology
Delaware	Medical Examiner	Appointed	Unspecified	
District of Columbia	Medical Examiner	Appointed	Board-certified forensic pathologist	
Florida	Medical Examiner	Appointed	Practicing physician in pathology	
Georgia	Coroner	Elected for four-year term	The mayor serves as coroner in municipalities with populations of 5,000 or less Complete basic training course provided by the Georgia Police Academy within six months of assuming office Complete a training course approved by the Georgia Coroner's Training Council on an annual basis	US citizen At least 25 years old Residence in the county of service Registered voter High school diploma
	Medical Examiner	Appointed	Licensed physician	Licensed to practice in the state At least one year of medical-legal training
Hawaii	Coroner	Appointed	The chief of police of the counties of Hawaii, Maui, and Kauai is the coroner	
	Medical Examiner	Appointed	Unspecified	
Idaho	Coroner	Appointed or elected for four-year term	Attend coroner's school within 12 months of assuming office Complete 24 hours of in-service training on a biennial calendar basis	US citizen Residence in the county of service for at least one year prior to assuming office

State	Death Investigators	Elected or Appointed	Qualifications	Other requirements
Illinois	Coroner	Appointed or elected for four-year term	Complete Illinois Law Enforcement Training Standards Board coroners training program within seven months of assuming office Complete 24 hours of accredited continuing education for coroners on an annual basis	
	Medical Examiner	Appointed: Only Cook County has a medical examiner	Unspecified	
Indiana	Coroner	Elected for four-year term	Complete 40 hours of a training course for coroners within six months of assuming office Attend eight hours of a training course for coroners on an annual basis	Residence in the county of service
Iowa	State Medical Examiner	Appointed	Board-certified forensic pathologist	Licensed to practice in the state
	County Medical Examiner	Appointed for two-year term	Licensed physician	Licensed to practice in the state
Kansas	Coroner	Appointed	Licensed physician	Licensed to practice in the state
Kentucky	Coroner	Elected for four-year term	Complete a 40-hour basic coroner course Complete a 16-hour coroner course on an annual basis	At least 24 years old Residence in the state for two years Residence in the county of service for one year immediately preceding service
	Medical Examiner	Appointed	Board-certified forensic pathologist	
Louisiana	Coroner	Elected for four-year term	Licensed physician Note: This requirement is waived in parishes without a qualified licensed physician.	Licensed to practice in the state Residence in the parish of service
Maine	Medical Examiner	Appointed	Licensed physician with expertise in forensic pathology	Licensed to practice in the state
Maryland	Medical Examiner	Appointed	Board-certified anatomic and forensic pathologist	
Massachusetts	Medical Examiner	Appointed	Board-certified anatomic and forensic pathologist	Licensed to practice in the state Residence in the state within six months of assuming position

State	Death Investigators	Elected or Appointed	Qualifications	Other requirements
Michigan	Medical Examiner	Appointed	Licensed physician	Licensed to practice in the state Note: If the county does not have an accredited hospital, licensed in another state that borders the county.
Minnesota	Coroner	Appointed or elected for four-year term	Licensed physician with training in medical-legal death investigation	
	Medical Examiner	Appointed	Board-certified forensic pathologist	
Mississippi	Coroner	Elected for four-year term	Complete the Mississippi Crime Laboratory and State Medical Examiner Death Investigation Training School course	At least 21 years old High School diploma Registered voter in the county of service
	State Medical Examiner	Appointed	Board-certified forensic pathologist	
	County Medical Examiner	Appointed	Licensed physician	
Missouri	Coroner	Elected for four-year term		US citizen At least 21 years old Residence in the state for one year, and within the county of service, six months immediately preceding service
	Medical Examiner	Appointed	Licensed physician	Licensed to practice in the state
Montana	Coroner	Appointed or elected	Complete the 40-hour basic coroner course taught by the Montana Public Safety Officer Standards and Training Council Complete 16-hour advanced coroner course on a biennial calendar basis	US citizen At least 18 years old High school diploma or equivalent Registered voter in the county of service
	Medical Examiner	Appointed	Licensed physician	Licensed to practice in the state
Nebraska	Coroner	Appointed or elected	Complete death investigation training within one year of service Complete continuing education as appropriate	
Nevada	Coroner	Appointed or elected for four-year term	Elected sheriff The boards of county commissioners may also appoint coroners when necessary	
New Hampshire	Medical Examiner	Appointed	Board-certified forensic pathologist	Licensed to practice in the state

State	Death Investigators	Elected or Appointed	Qualifications	Other requirements
New Jersey	State Medical Examiner	Appointed	Board-certified forensic pathologist	Licensed to practice in the state
	County Medical Examiner	Appointed	Licensed physician	Residence in the county of service Licensed to practice in the state Complete 30 hours of basic education in death investigation Complete basic course on the laws, rules, and regulations relating to the New Jersey Medical Examiner System Complete seven full days of internship training at the New Jersey State Medical Examiner Office
New Mexico	Medical Examiner	Appointed	Licensed physician with expertise in forensic pathology	Licensed to practice in the state
New York	Coroner	Elected for four-year term	Complete an introductory course in medical examination	At least 18 years old Residence in the county of service
	Medical Examiner	Appointed	Licensed physician	Licensed to practice in the state Residence in the county of service
North Carolina	Coroner	Elected for four-year term		Registered voter
	Chief Medical Examiner	Appointed	Board-certified forensic pathologist	Licensed to practice in the state
	County Medical Examiner	Appointed for three-year term	Licensed physician, licensed physician assistant, nurse practitioner, nurse, or emergency medical technician paramedic	Licensed to practice in the state
North Dakota	Coroner	Appointed for five-year term	Licensed physician, licensed physician assistant, licensed nurse, or any other individual determined by the state medical examiner to be qualified to serve as coroner	
	Medical Examiner	Appointed	Board-certified forensic pathologist	Licensed to practice in the state

State	Death Investigators	Elected or Appointed	Qualifications	Other requirements
Ohio	Coroner	Elected for four-year term	Licensed physician Complete 16 hours of continuing education sponsored by the Ohio State Coroners Association before assuming office Complete 32 hours of continuing education at programs sponsored by the Ohio State Coroners Association over the course of term	Licensed to practice in the state
	Medical Examiner	Appointed in Cuyahoga and Summit Counties	Licensed physician with expertise in forensic pathology	Licensed to practice in the state
Oklahoma	Chief Medical Examiner	Appointed	Board-certified forensic pathologist	Licensed to practice in the state
	County Medical Examiner	Appointed	Licensed physician	Licensed to practice in the state
Oregon	Medical Examiner	Appointed	State Medical Examiner Advisory Board recommends qualification of medical examiner	
Pennsylvania	Coroner	Elected for four-year term	Complete 32-hour course for new coroners Complete eight hours of continuing education on an annual basis	US citizen At least 18 years old Residence in the county for at least one year immediately preceding service
	Medical Examiner	Appointed in Allegheny, Delaware, and Philadelphia counties	Board-certified forensic pathologist	Licensed to practice in the state
Rhode Island	Medical Examiner	Appointed	Board-certified forensic pathologist	Licensed to practice in the state

State	Death Investigators	Elected or Appointed	Qualifications	Other requirements
South Carolina	Coroner	Elected for four-year term	Complete a basic training session no later than one year after assuming office. Complete 16 hours of continuing education on an annual basis	US citizen At least 21 years old High school diploma or equivalent Registered voter Residence in the county for at least one year immediately preceding service In addition to the above requirements, a coroner must have at least one of the following qualifications: (1) three years of experience in death investigation; (2) two-year associate degree and two years of experience in death investigation; (3) four-year baccalaureate degree and one year of experience in death investigation; (4) law enforcement officer; (5) completed a recognized forensic science degree or certification program; (6) medical doctor; or (7) have a bachelor of science degree in nursing
	Medical Examiner	Appointed	Licensed physician	
South Dakota	Coroner	Elected for four-year term	Complete 16-hour course sponsored by the Law Enforcement Officers Standards Commission within one year of assuming office. Complete eight hours of continuing education on a biennial basis	
	Medical Examiner	Appointed in counties with populations of 75,000 or more	Board-certified forensic pathologist	Licensed to practice in the state
Tennessee	Coroner	Elected for two-year term	licensed physician assistant, nurse, emergency medical technician paramedic, or diplomat of the American Board of Medicolegal Death Investigators	
	Chief Medical Examiner	Appointed	Board-certified forensic pathologist	Licensed to practice in the state
	County Medical Examiner	Appointed for four-year term	Licensed physician	Licensed to practice in the state

State	Death Investigators	Elected or Appointed	Qualifications	Other requirements
Texas	Coroner	Elected for four-year term	Justice of the Peace Complete a basic training session within first year of assuming position	
	Medical Examiner	Appointed	Licensed physician	Licensed to practice in the state
Utah	Medical Examiner	Appointed	Board-certified forensic pathologist	Licensed to practice in the state
Vermont	Medical Examiner	Appointed	Licensed physician	Licensed to practice in the state
Virginia	Medical Examiner	Appointed	Board-certified forensic pathologist	Licensed to practice in the state
Washington	Coroner	Elected for four-year term	Complete death investigation training course	
	Medical Examiner	Appointed	Board-certified forensic pathologist	Licensed to practice in the state
West Virginia	Coroner	Appointed	Licensed physician assistant, nurse, or emergency medical technician paramedic Complete a medical-legal death investigation course within one year of assuming position	
	Chief Medical Examiner	Appointed	Board-certified forensic pathologist	Licensed to practice in the state
	County Medical Examiner	Appointed for four-year term	Licensed physician, licensed physician assistant, nurse, or emergency medical technician paramedic Complete a medical-legal death investigation course within one year of assuming position	Licensed to practice in the state
Wisconsin	Coroner	Elected for four-year term	Unspecified	
	Medical Examiner	Appointed	Not required to be a physician	
Wyoming	Coroner	Elected for four-year term	Complete a basic coroner course within one year of service Complete continuing education as appropriate	

Source: www.cdc.gov/phlp/publications/topic/coroner.html

Acknowledgments

It was not an easy decision to divulge my drug use, even knowing that I was doing so as an act of civil disobedience. Had it not been for the support and encouragement of some profoundly humane individuals, *Drug Use for Grown Ups* would have remained in the closet, unavailable for public viewing.

I am indebted to Scott Moyers, my editor, who pushed me to write a book that would test my own integrity while confronting some of the most basic notions people believe about drugs and users. Your patient guidance, drug knowledge, and humble brilliance were precisely what I needed in order to see this project through to completion. Mia Council, super-efficient assistant editor, and Plaegian Alexander, copyeditor extraordinaire, thanks for getting me over the finish line.

Sascha Alper and Larry Weissman, my nurturing agents, deserve special gratitude. They championed this book from the very beginning and never wavered in their support, even in the face of an extended writing hiatus caused by my ill-advised decision to serve as department chairman. I owe thanks and praises to Claire Wachtel for connecting Larry and me, and for being a blunt and honest friend.

Much of the information presented throughout these pages was collected during my travels around the globe. Special thanks go to the numerous people who shared their homes, knowledge, and bounty. Many of you served as my therapist and restored my faith in humanity. I am especially grateful to Pat O'Hare, Barbara Boers, Anne Francois, Fabrice Olivet, Sebastian Saville, Lynne Lyman, Carla Shedd, Rick Doblin, Amir Bar-Lev, Iain (Buff) Cameron, Chris Rintoul, Katy MacLeod, Kirsten Horsburgh, Liliana Galindo, Guy Jones, Inez Feria Jorge, Marc Grifell, Nuria Calzada, Mireia Ventura, Kasia Malinowska-Sempruch, Julita Lemgruber, Bruno Torturra, Leon Garcia, Maria-Goretti Loglo, Pam Lichty, Teri Krebs, Julian Quintero, Annahita Mahdavi, Gamjad Paungsawad, Michael Schneider, Donald MacPherson, and Dean Wilson. Also, I would be remiss if I did not acknowledge the Brocher Foundation in Geneva and the Open Society Foundation for funds used during writing retreats.

My academic home affords me the opportunity to learn from some of the most critical thinkers in neuropsychopharmacology. I owe a tremendous debt to my coauthors, colleagues, and students. A good number of the ideas shared here resulted directly from our interactions. I am deeply appreciative to Elias Dakwar, Christopher Medina-Kirchner, Kate Y. O'Malley, Tiesha Gregory, Kristen Gwynne, Susie Swithers, Charles Ksir, James Rose, Samantha Santoscoy, Valerie Fendt, and Lynn Paltrow. Some of these individuals even read and reacted to early drafts of the manuscript.

Finally, I would like to thank the members of my family for their unconditional love and support. Malakai, Damon, Alise, and Tabius, your comments on select portions of earlier drafts were invaluable; they often reminded me of why I do what I do. Robin, you read every single page of the draft manuscript—which I know is painstaking work. The final product is considerably more human because of you. I would never have completed this book without your help and steadfast encouragement, truth be told.

Dear Dexie, without your fuel-infused love and abiding joy, I'd be a bitter, selfish, unproductive person.

Notes

PROLOGUE: TIME TO GROW UP

1. Locke J (1690), *An Essay concerning Human Understanding*, 4 vols (London: Printed by Eliz. Holt, for Thomas Basset, at the George in Fleet Street, near St. Dunstan's Church).
2. Miller LL, Cornett TL (1978), "Marijuana: Dose Effects on Pulse Rate, Subjective Estimates of Intoxication, Free Recall and Recognition Memory," *Pharmacology, Biochemistry, and Behavior* 9: 573–77.
3. Zorumski CF, Mennerick S, Izumi Y (2014), "Acute and Chronic Effects of Ethanol on Learning-Related Synaptic Plasticity," *Alcohol* 48: 1–17; Mintzer MZ, Griffiths RR (1999), "Triazolam and Zolpidem: Effects on Human Memory and Attentional Processes," *Psychopharmacology* 144: 8–19; Roy-Byrne P, Fleishaker J, Arnett C, Dubach M, Stewart J, Radant A, Veith R, Graham M (1993), "Effects of Acute and Chronic Alprazolam Treatment on Cerebral Blood Flow, Memory, Sedation, and Plasma Catecholamines," *Neuropsychopharmacology* 8: 161–69.
4. Haney M, Gunderson EW, Rabkin J, Hart CL, Vosburg SK, Comer SD, Foltin RW (2007), "Dronabinol and Marijuana in HIV+ Marijuana Smokers: Caloric Intake, Mood, and Sleep," *Journal of Acquired Immune Deficiency Syndrome* 45: 54–554; Hill KP (2015), "Medical Marijuana for Treatment of Chronic Pain and Other Medical and Psychiatric Problems: A Clinical Review," *Journal of the American Medical Association* 313: 2474–83.
5. Edwards E, Bunting W, Garcia L (2013), *The War on Marijuana in Black and White*, American Civil Liberties Union report.
6. U.S. Sentencing Commission (2017), Datafile, USSCFY17, accessed November 13, 2019: www.ussc.gov/sites/default/files/pdf/research-and-publications/annual-reports-and-sourcebooks/2017/Table34.pdf.
7. Reports and Detailed Tables from the 2018 National Survey on Drug Use and Health (NSDUH), accessed on November 13, 2019: www.samhsa.gov/data/nsduh/reports-detailed-tables-2018-NSDUH.

8. Riley KJ (1997), *Crack, Powder Cocaine, and Heroin: Drug Purchase and Use Patterns in Six U.S. Cities,* accessed on November 13, 2019: www.ncjrs.gov /pdffiles/167265.pdf.

9. Elkind MS, Sciacca R, Boden-Albala B, Rundek T, Paik MC, Sacco RL (2006), "Moderate Alcohol Consumption Reduces Risk of Ischemic Stroke: The Northern Manhattan Study," *Stroke* 37: 13–19; Poli A, Marangoni F, Avogaro A, Barba G, Bellentani S, Bucci M, Cambieri R, Catapano AL, Costanzo S, Cricelli C, de Gaetano G, Di Castelnuovo A, Faggiano P, Fattirolli F, Fontana L, Forlani G, Frattini S, Giacco R, La Vecchia C, Lazzaretto L, Loffredo L, Lucchin L, Marelli G, Marrocco W, Minisola S, Musicco M, Novo S, Nozzoli C, Pelucchi C, Perri L, Pieralli F, Rizzoni D, Sterzi R, Vettor R, Violi F, Visioli F (2013), "Moderate Alcohol Use and Health: A Consensus Document," *Nutrition, Metabolism, and Cardiovascular Diseases* 23: 487–504; Lee SJ, Cho YJ, Kim JG, Ko Y, Hong KS, Park JM, Kang K, Park TH, Park SS, Lee KB, Cha JK, Kim DH, Lee J, Kim JT, Lee J, Lee JS, Jang MS, Han MK, Gorelick PB, Bae HJ; CRCS-5 Investigators (2015), "Moderate Alcohol Intake Reduces Risk of Ischemic Stroke in Korea," *Neurology* 85: 1950–56; Bell S, Daskalopoulou M, Rapsomaniki E, George J, Britton A, Bobak M, Casas JP, Dale CE, Denaxas S, Shah AD, Hemingway H (2017), "Association between Clinically Recorded Alcohol Consumption and Initial Presentation of 12 Cardiovascular Diseases: Population-Based Cohort Study Using Linked Health Records," *British Medical Journal* 356: 909; Costanzo S, de Gaetano G, Di Castelnuovo A, Djoussé L, Poli A, van Velden DP (2019), "Moderate Alcohol Consumption and Lower Total Mortality Risk: Justified Doubts or Established Facts?," *Nutrition, Metabolism, and Cardiovascular Diseases* 29: 1003–1008.

10. Anthony JC, Warner LA, Kessler RC (1994), "Comparative Epidemiology of Dependence on Tobacco, Alcohol, Controlled Substances, and Inhalants: Basic Findings from the National Comorbidity Survey," *Experimental and Clinical Psychopharmacology* 2: 244–68; Warner LA, Kessler RC, Hughes M, Anthony JC, Nelson CB (1995), "Prevalence and Correlates of Drug Use and Dependence in the United States. Results from the National Comorbidity Survey," *Archives of General Psychiatry* 52: 219–29; O'Brien MS, Anthony JC (2009), "Extra-medical Stimulant Dependence among Recent Initiates," *Drug and Alcohol Dependence* 104: 147–55; Substance Abuse and Mental Health Services Administration (2012), *Results from the 2011 National Survey on Drug Use and Health: Summary of National Findings,* NSDUH Series H—44, HHS publication no. (SMA), 12–4713 (Rockville, MD: SAMHSA, 2012); Csete J, Kamarulzaman A, Kazatchkine M, Altice F, Balicki M, Buxton J, Cepeda J, Comfort M, Goosby E, Gouldo J, Hart C, Kerr T, Lajous AM, Lewis S, Martin N, Mejia D, Camacho A, Mathieson D, Obot I, Ogunrombi A, Sherman S, Stone J, Vallath N, Vickerman P, Zabransky T, Beyrer C (2016), "Public Health and International Drug Policy," *The Lancet* 387: 1427–80; Santiago Rivera OJ, Havens JR, Parker MA, Anthony JC (2018), "Risk of Heroin Dependence in Newly Incident Heroin Users," *Journal of the American Medical Association Psychiatry* 75: 863–64.

11. Bakken K, Landheim AS, Vaglum P (2007), "Axis I and II Disorders as Long-Term Predictors of Mental Distress: A Six-Year Prospective Follow-Up of Substance-Dependent Patients," *BMC Psychiatry* 7: 29; Moore E, Mancuso SG, Slade T, Galletly C, Castle DJ (2012), "The Impact of Alcohol and Illicit Drugs on People with Psychosis: The Second Australian National Survey of Psychosis," *Australian and New Zealand Journal of Psychiatry* 46: 864–78; Tsemberis S, Kent D, Respress C (2012), "Housing Stability and Recovery among Chronically Homeless Persons with Co-occurring Disorders in Washington, DC," *American Journal of Public Health* 102: 13–6; Tolliver BK, Anton RF (2015), "Assessment and Treatment of Mood Disorders in the Context of Substance Abuse," *Dialogues in Clinical Neuroscience* 17: 181–90; Grant BF, Saha TD, Ruan WJ, Goldstein RB, Chou SP, Jung J, Zhang H, Smith SM, Pickering RP, Huang B, Hasin DS (2016), "Epidemiology of DSM-5 Drug Use Disorder: Results from the National Epidemiologic Survey on Alcohol and Related Conditions–III," *JAMA Psychiatry* 73: 39–47; Lee JO, Jones TM, Kosterman R, Rhew IC, Lovasi GS, Hill KG, Catalano RF, Hawkins JD (2017), "The Association of Unemployment from Age 21 to 33 with Substance Use Disorder Symptoms at Age 39: The Role of Childhood Neighborhood Characteristics," *Drug and Alcohol Dependence* 174: 1–8; Thern E, Ramstedt M, Svensson J (October 16, 2019), "Long-term Effects of Youth Unemployment on Alcohol-Related Morbidity," *Addiction*, doi: 10.1111/add.14838 [epub ahead of print].

I THE WAR ON US: HOW WE GOT IN THIS MESS

1. National Drug Control Budget (March 2019), *FY2020 Funding Highlights*, accessed November 13, 2019: www.whitehouse.gov/wp-content/uploads/2019/03/FY-20-Budget-Highlights.pdf.
2. Hogan HL, Walke R (June 1, 1987), CRS Report for Congress. *Federal Drug Control: President's Budget Request for Fiscal Year 1988*, accessed on November 13, 2019: www.everycrsreport.com/files/19870601_87-479GOV_865e7ff39h27164e727c5c60bb87d2c89334d7aa.pdf.
3. Shumlin P (2014), State of the State Speech, www.governing.com/topics/politics/gov-vermont-peter-shumlin-state-address.html.
4. U.S. Department of Health and Human Services, Public Health Service, *National Household Survey on Drug Abuse, 1993* (Rockville, MD: Substance Abuse and Mental Health Services Administration, 1995); Riley KJ (1997), *Crack, Powder Cocaine, and Heroin: Drug Purchase and Use Patterns in Six U.S. Cities*, accessed on November 13, 2019: www.ncjrs.gov/pdffiles/167265.pdf.
5. United States Sentencing Commission (USSC) (May 2002), *Special Report to the Congress—Cocaine and Federal Sentencing Policy*, accessed on November 13, 2019: www.ussc.gov/sites/default/files/pdf/news/congressional-testimony-and-reports/drug-topics/200205-rtc-cocaine-sentencing-policy/200205_Cocaine_and_Federal_Sentencing_Policy.pdf.

6. Drucker E (2002), "Population Impact of Mass Incarceration under New York's Rockefeller Drug Laws: An Analysis of Years of Life Lost," *Journal of Urban Health: Bulletin of the New York Academy of Medicine* 79: 1–10.

7. Schatz A. (July 1971), "The War Within: Portraits of Vietnam Veterans Fighting Heroin Addiction," *Life*, accessed on November 13, 2019: time.com /3878718/vietnam-veterans-heroin-addiction-treatment-photos/.

8. Nixon RM (June 17, 1971), Special Message to the Congress on Drug-Abuse Prevention and Control, accessed on November 13, 2019: www.presidency .ucsb.edu/ws/?pid=3048.

9. Ranzal E (March 5, 1971), "Mayor Seeks $9.2-Million for Methadone," *The New York Times*.

10. Hart CL, Hart MZ (2019), "Opioid Crisis: Another Mechanism Used to Perpetuate American Racism," *Cultural Diversity and Ethnic Minority Psychology* 25: 6–11.

11. U.S. Sentencing Commission (2017), Datafile, USSCFY17, accessed November 13, 2019: www.ussc.gov/sites/default/files/pdf/research-and-publications /annual-reports-and-sourcebooks/2017/Table34.pdf; Riley KJ (1997), *Crack, Powder Cocaine, and Heroin: Drug Purchase and Use Patterns in Six U.S. Cities*, accessed on November 13, 2019: www.ncjrs.gov/pdffiles/167265.pdf.

12. James Baldwin's December 10, 1986, speech can be accessed at www.loc.gov /rr/record/pressclub/baldwin.html.

13. Baldwin J (1985), *The Price of the Ticket: Collected Nonfiction, 1948–1985* (New York: St. Martin's).

14. Hart CL, Ksir C (2018), *Drugs, Society, and Human Behavior*, 17th ed. McGraw-Hill: New York.

15. Holmes JM (1997), *Thomas Jefferson Treats Himself: Herbs, Physicke, and Nutrition in Early America* (Fort Valley, VA: Loft Press).

16. Hart CL, Ksir C (2018), *Drugs, Society, and Human Behavior*, 17th ed. (New York: McGraw-Hill).

17. Kane HH (1882), *Opium Smoking in America and China: A Study of Its Prevalence and Effects, Immediate and Remote, on the Individual and the Nation* (New York: G.P. Putnam's Sons).

18. Williams EH (February 8, 1914), "Negro Cocaine Fiends Are a New Southern Menace." *The New York Times*.

19. Substance Abuse and Mental Health Services Administration (2014), *The DAWN Report: Benzodiazepines in Combination with Opioid Pain Relievers or Alcohol: Greater Risk of More Serious ED Visit Outcomes* (Rockville, MD: SAMHSA), www.samhsa.gov/data/sites/default/files/DAWN-SR192-Benzo Combos-2014/DAWN-SR192-BenzoCombos-2014.pdf.

20. Santiago Rivera OJ, Havens JR, Parker MA, Anthony JC (2018), "Risk of Heroin Dependence in Newly Incident Heroin Users," *Journal of the American Medical Association Psychiatry*, 75: 863–64; Edlund MJ, Martin BC, Russo JE, DeVries A, Braden JB, Sullivan MD (2014), "The Role of Opioid Prescription in Incident Opioid Abuse and Dependence among Individuals with Chronic Noncancer Pain: The Role of Opioid Prescription," *Clinical*

Journal of Pain, 30: 557–64; Noble M, Treadwell JR, Tregear SJ, Coates VH, Wiffen PJ, Akafomo C, Schoelles KM (2010), "Long-term Opioid Management for Chronic Noncancer Pain," *Cochrane Database Systematic Review* 20, accessed on December 24, 2019: www.ncbi.nlm.nih.gov/pmc/articles/PMC 6494200/.

21. King S (1991), *The Dark Tower III: The Waste Lands* (Hampton Falls, NH: Donald M. Grant).

22. Goldensohn, R. (May 25, 2018), "They Shared Drugs. Someone Died. Does That Make Them Killers?" *The New York Times*, accessed on November 13, 2019: www.nytimes.com/2018/05/25/us/drug-overdose-prosecution-crime.html.

23. Gladwell M (January 14, 2019), "Is Marijuana as Safe as We Think? Permitting Pot Is One Thing; Promoting Its Use Is Another," *The New Yorker*.

24. Former governor Paul LePage's remarks were made at a Town Hall Form in 2016 and can be viewed at www.cnn.com/videos/us/2016/01/08/maine -governor-paul-lepage-shifty-d-money-drugs-sot.wmtw.

25. U.S. Sentencing Commission (2017), Datafile, USSCFY17, accessed on November 13, 2019: www.ussc.gov/sites/default/files/pdf/research-and-publications /annual-reports-and-sourcebooks/2017/Table34.pdf; Martins SS, Sarvet A, Santaella-Tenorio J, Saha T, Grant BF, Hasin DS (2017), "Changes in U.S. Lifetime Heroin Use and Heroin Use Disorder: Prevalence from 2001–2002 to 2012–2013," National Epidemiologic Survey on Alcohol and Related Conditions, *Journal of the American Medical Association Psychiatry*, 74: 445–55.

26. From unpublished writings of James Baldwin presented in the 2016 documentary *I Am Not Your Negro*.

2 GET OUT OF THE CLOSET: STOP BEHAVING LIKE CHILDREN

1. Greenwood M (January 4, 2018), "Bernie Sanders: Marijuana Isn't Heroin," *The Hill*, accessed on November 13, 2019: thehill.com/homenews/senate /367422-bernie-sanders-marijuana-isnt-heroin.

2. Gramlich J (August 16, 2019), *What the Data Says about Gun Deaths in the U.S.*, Pew Research Center Report, accessed on November 13, 2019: www .pewresearch.org/fact-tank/2019/08/16/what-the-data-says-about-gun -deaths-in-the-u-s/.

3. Scholl L, Seth P, Kariisa M, Wilson N, Baldwin G (2019), *Drug and Opioid-Involved Overdose Deaths—United States, 2013–2017*, Morbidity and Mortality Weekly Report 67: 1419–27: dx.doi.org/10.15585/mmwr.mm675152e1.

4. Occupational Health and Safety (February 18, 2019), *2018 Third Consecutive Year of at Least 40,000 Motor Vehicle Deaths*, accessed on November 13, 2019: ohsonline.com/Articles/2019/02/18/NSC-Motor-Vehicle-Deaths.aspx?Page=1.

5. Centers for Disease Control and Prevention (CDC) (January 3, 2018), Fact Sheets: *Alcohol Use and Health—Alcohol*, accessed on November 13, 2019: www.cdc.gov/alcohol/fact-sheets/alcohol-use.htm.

6. Kristof N (September 22, 2017), "How to Win a War on Drugs: Portugal Treats Addiction as a Disease, Not a Crime," *The New York Times*, accessed

on November 13, 2019: www.nytimes.com/2017/09/22/opinion/sunday /portugal-drug-decriminalization.html.

7. Staff writer (January 7, 1972), "Bush is Dead at Age 51, Was Psychology Professor," *Columbia Daily Spectator* CXVI, no. 49.

3 BEYOND THE HARMS OF HARM REDUCTION

1. Keneally M (June 27, 2019), "Opioids Responsible for Two-thirds of Global Drug Deaths in 2017: UN," *ABC News*. https://abcnews.go.com/International /opioids-responsible-thirds-global-drug-deaths-2017/story?id=63987167.
2. Jalal H, Buchanich JM, Roberts MS, Balmert LC, Zhang K, Burke DS (2018), "Changing Dynamics of the Drug Overdose Epidemic in the United States from 1979 through 2016," *Science* 361: 6408.
3. White AM, Castle IP, Hingson RW, Powell PA (2020), "Using Death Certificates to Explore Changes in Alcohol-Related Mortality in the United States, 1999 to 2017," *Alcoholism: Clinical and Experimental Research* 44: 178–87.
4. Van der Schrier R, Roozekrans M, Olofsen E, Aarts L, van Velzen M, de Jong M, Dahan A, Niesters M (2017), "Influence of Ethanol on Oxycodone-induced Respiratory Depression: A Dose-escalating Study in Young and Elderly Individuals," *Anesthesiology* 126: 534–42.
5. Klein D (August 8, 2018), "Mother Shocked as Task Force Recovers Enough Fentanyl to Kill 32,000 People," WSAZ News. www.wsaz.com/content/news /Man-arrested-on-drug-charges-in-Grayson-Ky-490416851.html.
6. Chason R (July 26, 2018), "Fentanyl-Related Deaths Continue 'Staggering' Rise in Maryland," *The Washington Post*. www.washingtonpost.com/local/md -politics/fentanyl-related-deaths-continue-staggering-rise-in-maryland /2018/07/26/cd33f406-90fc-11e8-8322-b5482bf5e0f5_story.html
7. Healy M (March 20, 2019), "Fentanyl Overdose Deaths in the U.S. Have Been Doubling Every Year," *The Los Angeles Times*. www.latimes.com/science /sciencenow/la-sci-sn-fentanyl-overdose-deaths-skyrocketing-2019 0320-story.html.
8. President Rodrigo Duterte (August 11, 2017), speech given in Davao City, Philippines, www.youtube.com/watch?v=qq_P3Yx8NAs.
9. Barratt M, Kowalski M, Maier L, Ritter A (2018), *Global Review of Drug Checking Services Operating in 2017*, Drug Policy Modelling Program Bulletin, no. 24 (Sydney: National Drug & Alcohol Research Centre, UNSW).
10. Goldmacher S (September 14, 2019), "Planned Parenthood and Fired Former Chief Mired in Escalating Dispute," *The New York Times*. www.nytimes.com /2019/09/14/us/politics/planned-parenthood-leana-wen.html.
11. Hart CL, Ksir C (2018), *Drugs, Society, and Human Behavior*, 17th ed. (New York: McGraw-Hill).
12. Monnat SM (2018), "Factors Associated with County-Level Differences in U.S. Drug-Related Mortality Rates, *American Journal of Preventive Medicine* 54: 611–19.

13. Schatz E, Nougier M (June 2012), *Drug Consumption Rooms: Evidence and Practice*, report from the International Drug Policy Consortium, accessed on December 24, 2019: fileserver.idpc.net/library/IDPC-Briefing-Paper_Drug -consumption-rooms.pdf.
14. Lit L, Schweitzer JB, Oberbauer AM (2011), "Handler Beliefs Affect Scent-Detection–Dog Outcomes," *Animal Cognition* 14: 387–94.

4 DRUG ADDICTION IS NOT A BRAIN DISEASE

1. Hart CL, Marvin CB, Silver R, Smith EE (2012), "Is Cognitive Functioning Impaired in Methamphetamine Users? A Critical Review," *Neuropsychopharmacology* 37: 586–608.
2. Leshner AI (1997), "Addiction Is a Brain Disease, and It Matters," *Science* 278: 45–47.
3. Hart CL, Jatlow PI, Sevarino KA, McCance-Katz EF (2000), "Comparison of Intravenous Cocaethylene and Cocaine in Humans," *Psychopharmacology* 149: 153–62.
4. United States Department of Health and Human Services, website retrieved on January 10, 2020, www.hhs.gov/programs/prevention-and-wellness/mental -health-substance-abuse/index.html.
5. Wallis C (April 19, 2019), "Pain Patients Get Relief from War on Opioids," *Scientific American*. www.scientificamerican.com/article/pain-patients-get -relief-from-war-on-opioids1/.
6. Salaverria LB (May 10, 2017), "Duterte Insists Shabu Can Cause Brain Damage," *Philippine Daily Inquirer*, accessed on November 13, 2019: newsinfo .inquirer.net/895885/duterte-insists-shabu-can-cause-brain-damage.
7. Volkow ND, Koob GF, McLellan AT (2016), "Neurobiologic Advances from the Brain-Disease Model of Addiction," *The New England Journal of Medicine* 374: 363–71.
8. American Psychiatric Association (2013), *Diagnostic and Statistical Manual of Mental Disorders*, 5th ed. (American Psychiatric Association: Washington, DC), 483.
9. Hart, CL (2017), "Viewing Addiction as a Brain Disease Promotes Social Injustice," *Nature: Human Behaviour*, www.nature.com/articles/s41562-017 -0055; Hart CL (2017), "Reply to: 'Addiction as a Brain Disease Does Not Promote Injustice,'" *Nature: Human Behaviour* 1: 611, www.nature.com /articles/s41562-017-0216-0.epdf; Grifell M, Hart, CL (2018), "Is Drug Addiction a Brain Disease? This Popular Claim Lacks Evidence and Leads to Poor Policy," *American Scientist* May–June, 160–67.
10. Gilman JM, Kuster JK, Lee S, Lee MJ, Kim BW, Makris N, van der Kouwe A, Blood AJ, Breiter HC (2014), "Cannabis Use Is Quantitatively Associated with Nucleus Accumbens and Amygdala Abnormalities in Young Adult Recreational Users," *Journal of Neuroscience* 34: 5529–38.
11. Schweitzer JB, Riggins T, Liang X, Gallen C, Kurup PK, Ross TJ, Black MM, Nair P, Salmeron BJ (2015), "Prenatal Drug Exposure to Illicit Drugs Alters

Working Memory–Related Brain Activity and Underlying Network Properties in Adolescence," *Neurotoxicology and Teratology* 48: 69–77.

12. McAllister D, Hart CL (2015), "Inappropriate Interpretations of Prenatal Drug Use Data Can Be Worse Than the Drugs Themselves," *Neurotoxicology and Teratology* 52 (Pt A): 57.

13. Johanson CE, Frey KA, Lundahl LH, Keenan P, Lockhart N, Roll J, Galloway GP, Koeppe RA, Kilbourn MR, Robbins T, Schuster CR (2006), "Cognitive Function and Nigrostriatal Markers in Abstinent Methamphetamine Abusers," *Psychopharmacology* 185: 327–38.

5 AMPHETAMINES: EMPATHY, ENERGY, AND ECSTASY

1. Walmsley R. (2015), *World Female Imprisonment List, 3rd ed: Women and Girls in Penal Institutions, including Pre-trial Detainees/Remand Prisoners* (Institute for Criminal Policy Research, Birkbeck, University of London), 1–15.

2. Jeffries S, Chuenurah C (2015), "Gender and Imprisonment in Thailand: Exploring the Trends and Understanding the Drivers," *International Journal of Law Crime and Justice* 45: 1–28.

3. Vongchak T, Kawichai S, Sherman S, Celentano D D, Sirisanthana T, Latkin C, Wiboonnatakul K, Srirak N, Jittiwutikarn J, Aramrattana A (2005), "The Influence of Thailand's 2003 'War on Drugs' Policy on Self-reported Drug Use among Injection Drug Users in Chiang Mai, Thailand," *International Journal of Drug Policy* 16: 115–21; United Nations Office on Drugs and Crime Regional Centre for East Asia and the Pacific (UNODC) (2007), *Patterns and Trends of Amphetamine-type Stimulants (ATS) and Other Drugs of Abuse in East Asia and the Pacific 2006: A Report from Project: TDRASF97 Improving ATS Data and Information Systems*, publication no. 2: 121–28.

4. Salaverria LB (May 10, 2017), "Duterte Insists Shabu Can Cause Brain Damage," *Philippine Daily Inquirer*, accessed on November 13, 2019: newsinfo.inquirer.net/895885/duterte-insists-shabu-can-cause-brain-damage.

5. Bueza M (2017), "In Numbers: The Philippines' 'War on Drugs'," *Rappler*. The figures update regularly on *Rappler*. Amnesty International last accessed the webpage on January 21, 2017.

6. Kirkpatrick MG, Gunderson EW, Johanson CE, Levin FR, Foltin RW, Hart CL (2012), "Comparison of Intranasal Methamphetamine and *d*-Amphetamine Self-Administration by Humans," *Addiction* 107: 783–91.

7. Kirkpatrick, Gunderson, Johanson, et al, "Comparison of Intranasal Methamphetamine," 783–91.

8. Hart CL, Ward AS, Haney M, Nasser J, Foltin RW (2003), "Methamphetamine Attenuates Disruptions in Performance and Mood During Simulated Night Shift Work." *Psychopharmacology* 169: 42–51.

9. Caldwell JA, Caldwell JL (2005), "Fatigue in Military Aviation: An Overview of U.S. Military-Approved Pharmacological Countermeasures," *Aviation Space & Environmental Medicine* 76: C39–51.

10. Hart CL, Ward AS, Haney M, Foltin RW, Fischman MW (2001), "Methamphetamine Self-administration by Humans," *Psychopharmacology* 157: 75–81.
11. Kirkpatrick, Gunderson, Johanson, et al, "Comparison of Intranasal Methamphetamine," 783–91.
12. Kirkpatrick MG, Gunderson EW, Perez AY, Haney M, Foltin RW, Hart CL (2012), "A Direct Comparison of the Behavioral and Physiological Effects of Methamphetamine and 3,4-Methylenedioxymethamphetamine (MDMA) in Humans," *Psychopharmacology* 219: 109–22.

6 NOVEL PSYCHOACTIVE SUBSTANCES: SEARCHING FOR A PURE BLISS

1. Papaseit E, Pérez-Mañá C, Mateus JA, Pujadas M, Fonseca F, Torrens M, Olesti E, de la Torre R, Farré M (2016), "Human Pharmacology of Mephedrone in Comparison with MDMA," *Neuropsychopharmacology* 41: 2704–13.
2. Luscombe R (May 29, 2012), "Miami Man Shot Dead Eating a Man's Face May Have Been on LSD-like Drug," *The Guardian*, accessed on November 14, 2019: www.theguardian.com/world/2012/may/29/miami-man-eating-face-lsd.
3. Staff writer (June 7, 2012), "New 'Bath Salts' Zombie-drug Makes Americans Eat Each Other," *RT*. www.rt.com/usa/drug-bath-salt-zombie-321/; Tienabeso S (May 29, 2012), "Face-Eating Attack Possibly Prompted by 'Bath Salts,' Authorities Suspect," *ABC News*. https://abcnews.go.com/US/face-eating-attack-possibly-linked-bath-salts-miami/story?id=16451452.
4. Swalve N, DeFoster R (2016), "Framing the Danger of Designer Drugs: Mass Media, Bath Salts, and the 'Miami Zombie Attack'," *Contemporary Drug Problems* 43: 103–21.
5. Firger J (April 2, 2015), "What Is Flakka? Florida's Dangerous New Drug Trend," *CBS News*, accessed on November 14, 2019: www.cbsnews.com/news/flakka-floridas-dangerous-new-drug-trend/.
6. Sullum J (November 13, 2017), "The Legend of Zombie Drugs Will Not Die," *Reason*, accessed on November 14, 2019: reason.com/2017/11/13/the-legend-of-zombie-drugs-will-not-die/.
7. Southall A, Ferré-Sadurní L (May 20, 2018), "K2 Eyed as Culprit after 14 People Overdose in Brooklyn," *The New York Times*, accessed on November 14, 2019: www.nytimes.com/2018/05/20/nyregion/k2-drug-overdose-brooklyn.html.
8. Miller, M. (July 13, 2016), "Synthetic Marijuana Overdose Turns Dozens into 'Zombies' in NYC," *CBS News*, accessed on November 14, 2019: www.cbsnews.com/news/synthetic-marijuana-overdose-turn-dozens-into-zombies-in-nyc/.
9. Miller M. "Synthetic Marijuana Overdose."
10. Rosenberg E, Schweber N (July 12, 2016), "33 Suspected of Overdosing on Synthetic Marijuana in Brooklyn," *The New York Times*, www.nytimes.com/2016/07/13/nyregion/k2-synthetic-marijuana-overdose-in-brooklyn.html.

11. Santora M. (December 14, 2016), "Drug 85 Times as Potent as Marijuana Caused a 'Zombielike' State in Brooklyn," *The New York Times*, www.nytimes .com/2016/12/14/nyregion/zombielike-state-was-caused-by-synthetic -marijuana.html.

12. Adams AJ, Banister SD, Irizarry L, Trecki J, Schwartz M, Gerona R (2017), "'Zombie' Outbreak Caused by the Synthetic Cannabinoid AMB-FUBINACA in New York," *The New England Journal of Medicine* 376: 235–42.

13. St. Pierre A (September 15, 2011), "Oh, the Irony: Speaker of the House John Boehner Continues to Support Marijuana Prohibition," *NORML Blog*, accessed on November 14, 2019: blog.norml.org/2011/09/15/oh-the-irony -speaker-of-the-house-john-boehner-continues-to-support-marijuana -prohibition/.

14. Breslow J (March 16, 2019), "John Boehner Was Once 'Unalterably Opposed' to Marijuana. He Now Wants It To Be Legal," *National Public Radio*, accessed on November 14, 2019: www.npr.org/2019/03/16/704086782/john-boehner -was-once-unalterably-opposed-to-marijuana-he-now-wants-it-to-be-leg.

15. Assari S, Moazen-Zadeh E, Caldwell CH, Zimmerman MA (2017), "Racial Discrimination during Adolescence Predicts Mental Health Deterioration in Adulthood: Gender Differences among Blacks," *Frontiers in Public Health* 5: 104.

16. Dolezsar CM, McGrath JJ, Herzig AJ, Miller SB (2014), "Perceived Racial Discrimination and Hypertension: A Comprehensive Systematic Review," *Health Psychology* 33: 20–34.

7 CANNABIS: SPROUTING THE SEEDS OF FREEDOM

1. Transcript of Testimony Heard by the Grand Jury Charged with Determining Whether Police Officer Darren Wilson Would Be Indicted for Killing Michael Brown. https://edition.cnn.com/interactive/2014/11/us/ferguson-grand -jury-docs/index.html

2. Staff writer (September 24, 2016), "Charlotte Police Release Official Footage of Fatal Keith Lamont Scott Shooting," *Complex*, www.complex.com/life /2016/09/charlotte-police-release-keith-scott-footage.

3. Hart CL (July 11, 2013), "Reefer Madness, an Unfortunate Redux," *The New York Times*: www.nytimes.com/2013/07/12/opinion/reefer-madness-an -unfortunate-redux.html?_r=0.

4. Hart CL, van Gorp WG, Haney M, Foltin RW, Fischman MW (2001), "Effects of Acute Smoked Marijuana on Complex Cognitive Performance," *Neuropsychopharmacology* 25: 757–65; Hart CL, Haney M, Ward AS, Fischman MW, Foltin RW (2002), "Effects of Oral THC Maintenance on Smoked Marijuana Self-Administration," *Drug and Alcohol Dependence* 67: 301–309; Hart CL, Ward AS, Haney M, Comer SD, Foltin RW, Fischman MW (2002), "Comparison of Smoked Marijuana and Oral Δ9-Tetrahydrocannabinol in Humans," *Psychopharmacology* 164: 407–15;

Hart CL, Ilan AB, Gevins A, Gunderson EW, Role K, Colley J, Foltin RW (2010), "Neurophysiological and Cognitive Effects of Smoked Marijuana in Frequent Users," *Pharmacology, Biochemistry, and Behavior* 96: 333–41; Keith DR, Gunderson EW, Haney M, Foltin RW, Hart CL (2017), "Smoked Marijuana Attenuates Performance Disruptions during Simulated Night Shift Work," *Drug and Alcohol Dependence* 178: 534–43.

5. Anslinger, H. J., & C. R. Cooper. (1937). "Marijuana: Assassin of Youth," *The American Magazine* 124: 19, 153.

6. Berenson A (2019), *Tell Your Children: The Truth about Marijuana, Mental Illness, and Violence* (New York: Free Press).

7. Associated Press (July 29, 2007), "Even Infrequent Use of Marijuana Increases Risk of Psychosis by 40 Percent," Fox News, www.foxnews.com/story /study-even-infrequent-use-of-marijuana-increases-risk-of-psychosis-by -40-percent.

8. Mustonen A, Ahokas T, Nordström T, Murray GK, Mäki P, Jääskeläinen E, Heiskala A, Mcgrath JJ, Scott JG, Miettunen J, Niemelä S (2018), "Smokin' Hot: Adolescent Smoking and the Risk of Psychosis," *Acta Psychiatrica Scandinavica* 138: 5–14.

9. Moran LV, Masters GA, Pingali S, Cohen BM, Liebson E, Rajarethinam RP, Ongur D (2015), "Prescription Stimulant Use Is Associated with Earlier Onset of Psychosis," *Journal of Psychiatric Research* 71: 41–47.

10. Torrey EF, Simmons W, Yolken RH (2015), "Childhood Cat Ownership a Risk Factor for Schizophrenia Later in Life?," *Schizophrenia Research* 165: 1–2.

11. Ksir C, Hart CL (2016), "Cannabis and Psychosis: A Critical Overview of the Relationship," *Current Psychiatry Reports* 18: 12; Ksir C, Hart CL (2016), "Correlation Still Does Not Imply Causation," *The Lancet* 3: 401.

12. Detailed Tables from the 2018 National Survey on Drug Use and Health (NSDUH), accessed on January 14, 2020: www.samhsa.gov/data/sites/default /files/cbhsq-reports/NSDUHDetailedTabs2018R2/NSDUHDetailedTabs 2018.pdf.

13. Hearing Before the Subcommittee on Government Operations of the Committee on Oversight and Government Reform House of Representatives, One Hundred Thirteenth Congress, Second Session. June 20, 2014. Starting at 08:10: www.youtube.com/watch?v=M6CSc4nl—Q&t=33s.

14. McLeod, E., Friedman, A., & Soderberg, B. (December 2018), "Structural Racism and Cannabis: Black Baltimoreans Still Disproportionately Arrested for Weed Decriminalization," *A Baltimore Fishbowl Report*, accessed on November 14, 2019: baltimorefishbowl.com/stories/structural-racism-and -cannabis-black-baltimoreans-still-disproportionately-arrested-for-weed -after-decriminalization/.

15. Hannon E (January 29, 2019), "Baltimore Will Stop Prosecuting Marijuana Possession Cases, as State's Attorney Moves to Vacate Thousands of Prior Convictions," *Slate.* https://slate.com/news-and-politics/2019/01/baltimore

-stop-prosecuting-marijuana-possession-cases-states-attorney-marilyn
-mosby-vacate-prior-convictions.html.

16. McCarthy J (October 22, 2018), "Two in Three Americans Now Support Legalizing Marijuana," *Gallup*, https://news.gallup.com/poll/243908/two-three-americans-support-legalizing-marijuana.aspx.

17. "Mayor LaGuardia's Committee on Marijuana," in D. Solomon, ed., *The Marihuana Papers* (New York: New American Library, 1966).

18. Detailed Tables from the 2019 Monitoring the Future Survey (MTF), accessed on January 14, 2020: www.monitoringthefuture.org/data/19data/19drtbl1.pdf.

19. Miech RA, Johnston LD, O'Malley PM, Bachman JG, Schulenberg JE, Patrick ME (2019), "Monitoring the Future National Survey Results on Drug Use, 1975–2018: Volume I, Secondary School Students," Ann Arbor: Institute for Social Research, The University of Michigan. www.monitoringthefuture.org//pubs/monographs/mtf-vol1_2018.pdf.

20. Information about marijuana sales and tax revenue retrieved from the Colorado Department of Revenue on October 24, 2019: www.colorado.gov/pacific/revenue/colorado-marijuana-tax-data; www.colorado.gov/pacific/revenue/colorado-marijuana-sales-reports.

21. Hopkins E (October 18, 1990), "Childhood's End: What Life Is Like for Crack Babies," *Rolling Stone*. www.rollingstone.com/culture/culture-news/childhoods-end-what-life-is-like-for-crack-babies-188557/.

22. Chasnoff IJ (2017), "Medical Marijuana Laws and Pregnancy: Implications for Public Health Policy," *American Journal of Obstetrics and Gynecology* 216: 27–30.

23. Torres CA, Hart CL (2017), "Marijuana and Pregnancy: Objective Education Is Good, Biased Education Is Not," *American Journal of Obstetrics and Gynecology* 217: 227; Torres CA, Medina-Kirchner C, O'Malley KY, Hart CL (2020), "Totality of the Evidence Suggest Prenatal Cannabis Exposure Does Not Lead to Cognitive Impairments: A Systematic and Critical Review," *Frontiers in Psychology* 11: 816 doi: 10.3389/fpsyg.2020.00816.

24. Doyle JJ (2008), "Child Protection and Adult Crime: Using Investigator Assignment to Estimate Causal Effects of Foster Care," *Journal of Political Economy* 116: 746–70.

25. Ko JY, Farr SL, Tong VT, Creanga AA, Callaghan WM (2015), "Prevalence and Patterns of Marijuana Use among Pregnant and Nonpregnant Women of Reproductive Age," *American Journal of Obstetrics and Gynecology* 213: 201.

26. Stadterman JM, Hart CL (2015), "Screening Women for Marijuana Use Does More Harm Than Good," *American Journal of Obstetrics and Gynecology* 213: 598–99.

8 PSYCHEDELICS: WE ARE ONE

1. Berman RM, Cappiello A, Anand A, Oren DA, Heninger GR, Charney DS, Krystal JH (2000), "Antidepressant Effects of Ketamine in Depressed

Patients," *Biological Psychiatry* 47: 351–54; Newport DJ, Carpenter LL, McDonald WM, Potash JB, Tohen M, Nemeroff CB (2015), "Ketamine and Other NMDA Antagonists: Early Clinical Trials and Possible Mechanisms in Depression," *American Journal of Psychiatry* 172: 950–66; Dakwar E, Levin F, Foltin RW, Nunes EV, Hart CL (2014), "The Effects of Sub-anesthetic Ketamine Infusions on Motivation to Quit and Cue-induced Craving in Cocaine Dependent Research Volunteers," *Biological Psychiatry* 76: 40–46; Griffiths RR, Johnson MW, Richards WA, Richards BD, Jesse R, MacLean KA, Barrett FS, Cosimano MP, Klinedinst MA (2018), "Psilocybin-occasioned Mystical-type Experience in Combination with Meditation and Other Spiritual Practices Produces Enduring Positive Changes in Psychological Functioning and in Trait Measures of Prosocial Attitudes and Behaviors," *Journal of Psychopharmacology* 32: 49–69; Johnson MW, Griffiths RR (2017), "Potential Therapeutic Effects of Psilocybin," *Neurotherapeutics* 14: 734–40.

2. Pollan M (2018), *How to Change Your Mind* (New York: Penguin).

3. Waldman A (2017), *A Really Good Day: How Microdosing Made a Mega Difference in My Mood, My Marriage, and My Life* (New York: Alfred A. Knopf).

4. Fadiman J (2011), *The Psychedelic Explorer's Guide: Safe, Therapeutic, and Sacred Journey* (Rochester, VT: Park Street Press).

5. Senate Communications Division (November 12, 1999), "Senator Asks Governor to Apologize for Racial Comments; Dickerson Calls Keating Statements Inappropriate, Offensive," accessed October 24, 2019: www.oksenate .gov/news/press_releases/press_releases_1999/PR991112.html.

6. Domino EF, Luby ED (2012), "Phencyclidine/Schizophrenia: One View toward the Past, the Other to the Future," *Schizophrenia Bulletin* 38: 914–19.

7. Brecher M, Wang BW, Wong H, Morgan JP (1988), "Phencyclidine and Violence: Clinical and Legal Issues," *Journal of Clinical Psychopharmacology* 8: 397–401.

8. Gorner J (April 15, 2015), "PCP Found in Body of Teen Shot 16 Times by Chicago Cop," *Chicago Tribune*.

9. Ford Q (October 21, 2014), "Cops: Boy, 17, Fatally Shot by Officer after Refusing to Drop Knife," *Chicago Tribune*.

10. Kalven J (February 10, 2015), "Sixteen Shots: Chicago Police Have Told Their Version of How 17-Year-Old Black Teen Laquan McDonald Died. The Autopsy Tells a Different Story," *Slate*, accessed on November 14, 2019: slate .com/news-and-politics/2015/02/laquan-mcdonald-shooting-a-recently -obtained-autopsy-report-on-the-dead-teen-complicates-the-chicago-police -departments-story.html.

11. Interview with Anita Alvarez in the documentary *16 Shots*.

9 COCAINE: EVERYBODY LOVES THE SUNSHINE

1. In January 2019, Jean Wyllys fled the country and gave up his congressional seat because of threats and fears for his life.

2. Boiteux L (2015), "Brazil: Critical Reflections on a Repressive Drug Policy,"

Sur International Journal of Human Rights 12; Izsák-Ndiaye R (February 9, 2016), *Report of the Special Rapporteur on Minority Issues on Her Mission to Brazil*, accessed on November 14, 2019: digitallibrary.un.org/record/831487?ln=en.

3. Morrison T (1997), *Paradise* (New York: Alfred A. Knopf).

4. Mena F (August 29, 2015), "Neurocientista negro diz ter sido barrado em hotel em SP. *Folha de S.Paulo*," accessed on November 14, 2019: www1.folha.uol.com.br/cotidiano/2015/08/1675340-neurocientista-negro-e-barrado-em-hotel-onde-ministraria-palestra-em-sp.shtml.

5. Frank DA, Augustyn M, Knight WG, Pell T, Zuckerman B (2001), "Growth, Development, and Behavior in Early Childhood Following Prenatal Cocaine Exposure: A Systematic Review," *Journal of the American Medical Association* 285: 1613–25; Frank DA, Jacobs RR, Beeghly M, Augustyn M, Bellinger D, Cabral H, Heeren T (2002), "Level of Prenatal Cocaine Exposure and Scores on the Bayley Scales of Infant Development: Modifying Effects of Caregiver, Early Intervention, and Birth Weight," *Pediatrics* 110: 1143–52; Beeghly M, Martin B, Rose-Jacobs R, Cabral H, Heeren T, Augustyn M, Bellinger D, Frank DA (2006), "Prenatal Cocaine Exposure and Children's Language Functioning at 6 and 9.5 Years: Moderating Effects of Child Age, Birthweight, and Gender," *Journal of Pediatric Psychology* 31: 98–115; Lewis BA, Minnes S, Short EJ, Weishampel P, Satayathum S, Min MO, Nelson S, Singer LT (2011), "The Effects of Prenatal Cocaine on Language Development at 10 Years of Age," *Neurotoxicology Teratology* 33: 17–24; Betancourt LM, Yang W, Brodsky NL, Gallagher PR, Malmud EK, Giannetta JM, Farah MJ, Hurt H (2011), "Adolescents with and without Gestational Cocaine Exposure: Longitudinal Analysis of Inhibitory Control, Memory and Receptive Language," *Neurotoxicology Teratology* 33: 36–46.

6. Cooper BM (December 1, 1987), "Kids Killing Kids: New Jack City Its Young," *The Village Voice*.

7. Ebert R (May 1, 1991), *New Jack City* movie review. www.rogerebert.com/reviews/new-jack-city-1991.

8. Anthony JC, Warner LA, Kessler RC (1994), "Comparative Epidemiology of Dependence on Tobacco, Alcohol, Controlled Substances, and Inhalants: Basic Findings from the National Comorbidity Survey," *Experimental and Clinical Psychopharmacology* 2: 244–68; Warner LA, Kessler RC, Hughes M, Anthony JC, Nelson CB (1995), "Prevalence and Correlates of Drug Use and Dependence in the United States. Results from the National Comorbidity Survey," *Archives of General Psychiatry* 52: 219–29; O'Brien MS, Anthony JC (2009), "Extra-medical Stimulant Dependence among Recent Initiates," *Drug and Alcohol Dependence* 104: 147–55; Substance Abuse and Mental Health Services Administration (2012), *Results from the 2011 National Survey on Drug Use and Health: Summary of National Findings*, NSDUH Series H—44, HHS publication no. (SMA) 12–4713 (Rockville, MD: SAMHSA); Csete J, Kamarulzaman A, Kazatchkine M, Altice F, Balicki M, Buxton J, Cepeda J, Comfort M, Goosby E, Gouldo J, Hart C, Kerr T, Lajous AM,

Lewis S, Martin N, Mejia, D, Camacho A, Mathieson D, Obot I, Ogunrombi A, Sherman S, Stone J, Vallath N, Vickerman P, Zabransky T, Beyrer C (2016), "Public Health and International Drug Policy," *The Lancet* 387: 1427–80; Santiago Rivera, OJ, Havens JR, Parker MA, Anthony JC (2018), "Risk of Heroin Dependence in Newly Incident Heroin Users," *Journal of the American Medical Association Psychiatry* 75: 863–64.

9. Hart CL, Haney M, Foltin RW, Fischman MW (2000), "Alternative Reinforcers Differentially Modify Cocaine Self-Administration by Humans," *Behavioural Pharmacology* 11: 87–91; Foltin RW, Ward AS, Haney M, Hart CL, Collins ED (2003), "The Effects of Escalating Doses of Smoked Cocaine in Humans," *Drug and Alcohol Dependence* 70: 149–57; Hart CL, Haney M, Collins ED, Rubin E, Foltin RW (2007), "Smoked Cocaine Self-Administration by Humans Is Not Reduced by Large Gabapentin Maintenance Doses," *Drug and Alcohol Dependence* 86: 274–77; Hart CL, Haney M, Vosburg SK, Rubin E, Foltin RW (2008), "Smoked Cocaine Self-Administration Is Decreased by Modafinil," *Neuropsychopharmacology* 33: 761–68.

10. Rosales K, Barnes T (September 14, 2011), "New Jack Rio," *Foreign Policy*, accessed on November 14, 2019: foreignpolicy.com/2011/09/14/new-jack-rio/.

11. Barbara V (May 22, 2018), "The Men Who Terrorize Rio," *The New York Times*. www.nytimes.com/2018/05/22/opinion/rio-janeiro-terrorize-militias.html.

12. *Por GloboNews* (August 28, 2018), "Rio de Janeiro Has an Average of Four Deaths per Day Caused by Police Intervention in 2018," g1.globo.com/rj/rio-de-janeiro/noticia/2018/08/28/rj-tem-media-de-4-mortes-por-dia-causadas-por-intervencao-policial-em-2018.ghtml.

13. Londoño E, Andreoni M (May 26, 2019), "'They Came to Kill': Almost 5 Die Daily at Hands of Rio Police," *The New York Times*. www.nytimes.com/2019/05/26/world/americas/brazil-rio-police-kill.html.

IO DOPE SCIENCE: THE TRUTH ABOUT OPIOIDS

1. Edlund MJ, Martin BC, Russo JE, DeVries A, Braden JB, Sullivan MD (2014), "The Role of Opioid Prescription in Incident Opioid Abuse and Dependence among Individuals with Chronic Noncancer Pain: The Role of Opioid Prescription," *Clinical Journal of Pain* 30: 557–64; Noble M, Treadwell JR, Tregear SJ, Coates VH, Wiffen PJ, Akafomo C, Schoelles KM (2010), "Long-term Opioid Management for Chronic Noncancer Pain," *Cochrane Database Systematic Review* 20, www.ncbi.nlm.nih.gov/pmc/articles/PMC6494200/.

2. Santiago Rivera, OJ, Havens JR, Parker MA, Anthony JC (2018), "Risk of Heroin Dependence in Newly Incident Heroin Users," *Journal of the American Medical Association Psychiatry* 75: 863–64.

3. Webster LR (2017), "Risk Factors for Opioid-Use Disorder and Overdose," *Anesthesia & Analgesia* 125: 1741–48.

4. Khan R, Khazaal Y, Thorens G, Zullino D, Uchtenhagen A (2014), "Understanding Swiss Drug-Policy Change and the Introduction of Heroin Maintenance Treatment," *European Addiction Research* 20: 200–207.
5. Stephenson C (August 11, 2016), "Heroin Suspected in 20 Deaths in 2 Weeks," *Milwaukee Journal Sentinel*. www.jsonline.com/story/news/crime/2016/08/11/fentanyl-deaths-spike/88580884/.
6. Yoon E, Babar A, Choudhary M, Kutner M, Pyrsopoulos N (2016), "Acetaminophen-Induced Hepatotoxicity: A Comprehensive Update," *Journal of Clinical and Translational Hepatology* 4: 131–42.
7. Serper M, Wolf MS, Parikh NA, Tillman H, Lee WM, Ganger DR (2016), "Risk Factors, Clinical Presentation, and Outcomes in Overdose with Acetaminophen Alone or with Combination Products: Results from the Acute Liver-Failure Study Group," *Journal of Clinical Gastroenterology* 50: 85–91.
8. Griffin FJ (August 23, 2019), "Returning to Lady: A Reflection on Two Decades 'In Search of Billie Holiday,'" *National Public Radio*, accessed on November 14, 2019: www.npr.org/2019/08/23/748740849/returning-to-lady-a-reflection-on-two-decades-in-search-of-billie-holiday.

EPILOGUE: THE JOURNEY

1. Substance Abuse and Mental Health Services Administration (2019), *Key Substance Use and Mental Health Indicators in the United States: Results from the 2018 National Survey on Drug Use and Health* (HHS publication no. PEP19-5068, NSDUH Series H-54) (Rockville, MD: Center for Behavioral Health Statistics and Quality, SAHMSA), retrieved from www.samhsa.gov/data/.
2. Clark NH (1965), *The Dry Years: Prohibition and Social Change in Washington* (Seattle: University of Washington Press).
3. Blum D (2010), *The Poisoner's Handbook: Murder and the Birth of Forensic Medicine in Jazz Age* (New York: Penguin).
4. Measham FC (2019), "Drug-safety Testing, Disposals and Dealing in an English Field: Exploring the Operational and Behavioural Outcomes of the UK's First Onsite 'Drug Checking' Service," *International Journal of Drug Policy* 67: 102–107.
5. Yoon E, Babar A, Choudhary M, Kutner M, Pyrsopoulos N (2016), "Acetaminophen-Induced Hepatotoxicity: A Comprehensive Update," *Journal of Clinical and Translational Hepatology* 4: 131–42.
6. Saad MH, Savonen CL, Rumschlag M, Todi SV, Schmidt CJ, Bannon MJ (2018), "Opioid Deaths: Trends, Biomarkers, and Potential Drug Interactions Revealed by Decision-Tree Analyses," *Frontiers in Neuroscience* 12: 728; Hopkins RE, Dobbin M, Pilgrim JL (2018), "Unintentional Mortality Associated with Paracetamol and Codeine Preparations, with and without Doxylamine, in Australia," *Forensic Science International* 282: 122–26.
7. King ML (April 16, 1963), "Letter from Birmingham Jail," http://web.cn.edu/kwheeler/documents/letter_birmingham_jail.pdf.

Index